Risk Assessment and Management

at

Deseret Chemical Depot

and the

Tooele Chemical Agent

Disposal Facility

Committee on Review and Evaluation of the
Army Chemical Stockpile Disposal Program

Board on Army Science and Technology

Commission on Engineering and Technical Systems

National Research Council

NATIONAL ACADEMY PRESS
Washington, D.C. 1997

NATIONAL ACADEMY PRESS • 2101 Constitution Avenue, N.W. • Washington, D.C. 20418

NOTICE: The project that is the subject of this report was approved by the Governing Board of the National Research Council, whose members are drawn from the councils of the National Academy of Sciences, the National Academy of Engineering, and the Institute of Medicine. The members of the committee responsible for the report were chosen for their special competences and with regard for appropriate balance.

This report has been reviewed by a group other than the authors according to procedures approved by a Report Review Committee consisting of members of the National Academy of Sciences, the National Academy of Engineering, and the Institute of Medicine.

The National Academy of Sciences is a private, nonprofit, self-perpetuating society of distinguished scholars engaged in scientific and engineering research, dedicated to the furtherance of science and technology and to their use for the general welfare. Upon the authority of the charter granted to it by the Congress in 1863, the Academy has a mandate that requires it to advise the federal government on scientific and technical matters. Dr. Bruce M. Alberts is president of the National Academy of Sciences.

The National Academy of Engineering was established in 1964, under the charter of the National Academy of Sciences, as a parallel organization of outstanding engineers. It is autonomous in its administration and in the selection of its members, sharing with the National Academy of Sciences the responsibility for advising the federal government. The National Academy of Engineering also sponsors engineering programs aimed at meeting national needs, encourages education and research, and recognizes the superior achievements of engineers. Dr. William A. Wulf is president of the National Academy of Engineering.

The Institute of Medicine was established in 1970 by the National Academy of Sciences to secure the services of eminent members of appropriate professions in the examination of policy matters pertaining to the health of the public. The Institute acts under the responsibility given to the National Academy of Sciences by its congressional charter to be an adviser to the federal government and, upon its own initiative, to identify issues of medical care, research, and education. Dr. Kenneth I. Shine is president of the Institute of Medicine.

The National Research Council was organized by the National Academy of Sciences in 1916 to associate the broad community of science and technology with the Academy's purposes of furthering knowledge and advising the federal government. Functioning in accordance with general policies determined by the Academy, the Council has become the principal operating agency of both the National Academy of Sciences and the National Academy of Engineering in providing services to the government, the public, and the scientific and engineering communities. The council is administered jointly by both Academies and the Institute of Medicine. Dr. Bruce M. Alberts and Dr. William A. Wulf are chairman and vice chairman, respectively, of the National Research Council.

This is a report of work supported by Contract DAAG55-97-C-0026 between the U.S. Army and the National Academy of Sciences.

Any opinions, findings, conclusions, or recommendations expressed in this publication are those of the author(s) and do not necessarily reflect the view of the organizations or agencies that provided support for the project.

International Standard Book Number 0-309-05841-4

Limited copies are available from: *Additional copies are available for sale from:*

Board on Army Science and Technology
National Research Council
2101 Constitution Avenue, N.W.
Washington, D.C. 20418
202-334-3918

National Academy Press
2101 Constitution Avenue, N.W.
Washington, D.C. 20418
800-624-6242 or
202-334-3313 (in the Washington Metropolitan Area)

Copyright 1997 by the National Academy of Sciences. All rights reserved.

Printed in the United States of America

COMMITTEE ON REVIEW AND EVALUATION OF THE ARMY CHEMICAL STOCKPILE DISPOSAL PROGRAM

RICHARD S. MAGEE (chair), New Jersey Institute of Technology, Newark
ELISABETH M. DRAKE (vice chair), Massachusetts Institute of Technology, Cambridge
DENNIS C. BLEY, Buttonwood Consulting, Inc., Oakton, Virginia
GENE H. DYER, consultant, San Rafael, California
VINCENT E. FALTER, U.S. Army (retired), Springfield, Virginia
J. ROBERT GIBSON, DuPont Agricultural Products, Wilmington, Delaware
MICHAEL R. GREENBERG, Rutgers, the State University of New Jersey, Piscataway
CHARLES E. KOLB, Aerodyne Research, Inc., Billerica, Massachusetts
DAVID S. KOSSON, Rutgers, the State University of New Jersey, Piscataway
WALTER G. MAY, University of Illinois, Urbana
ALVIN H. MUSHKATEL, Arizona State University, Tempe
PETER J. NIEMIEC, Greenberg, Glusker, Fields, Claman & Machtinger LLP, Los Angeles, California
GEORGE W. PARSHALL, DuPont Company (retired), Wilmington, Delaware
WILLIAM TUMAS, Los Alamos National Laboratory, Los Alamos, New Mexico
JYA-SYIN WU, Hughes Information Technology Systems, Fullerton, California

Board on Army Science and Technology Liaison

ROBERT A. BEAUDET, University of Southern California, Los Angeles

Staff

DONALD L. SIEBENALER, Study Director
HARRISON T. PANNELLA, Consultant
SHIREL R. SMITH, Senior Project Assistant

BOARD ON ARMY SCIENCE AND TECHNOLOGY

GENERAL GLENN K. OTIS (chair), U.S. Army (retired), Newport News, Virginia
CHRISTOPHER C. GREEN (vice chair) General Motors Corporation, Warren, Michigan
ROBERT A. BEAUDET, University of Southern California, Los Angeles
GARY L. BORMAN, University of Wisconsin, Madison
LAWRENCE J. DELANEY, BDM Europe, Berlin, Germany
WILLIAM H. FORSTER, Northrop Grumman Corporation, Baltimore, Maryland
ROBERT J. HEASTON, Guidance and Control Information Analysis Center (retired), Naperville, Illinois
KATHRYN V. LOGAN, Georgia Institute of Technology, Atlanta
THOMAS MCNAUGHER, RAND, Washington, D.C.
NORMAN F. PARKER, Varian Associates (retired), Cardiff by the Sea, California
STEWART D. PERSONICK, Bell Communications Research, Inc., Morristown, New Jersey
MILLARD "FRANK" ROSE, Auburn University, Auburn, Alabama
HARVEY W. SCHADLER, General Electric, Schenectady, New York
CLARENCE G. THORNTON, Army Research Laboratories (retired), Colts Neck, New Jersey
JOHN D. VENABLES, Martin Marietta Laboratories (retired), Towson, Maryland
ALLEN C. WARD, University of Michigan, Ann Arbor

Staff

BRUCE A. BRAUN, Director
MICHAEL A. CLARKE, Senior Program Officer
ROBERT J. LOVE, Senior Program Officer
DONALD L. SIEBENALER, Study Director
MARGO L. FRANCESCO, Staff Associate
ALVERA GIRCYS, Financial Assistant
JACQUELINE CAMPBELL-JOHNSON, Senior Project Assistant
CECELIA RAY, Senior Project Assistant
SHIREL R. SMITH, Senior Project Assistant

Preface

In 1985, Congress directed the U.S. Army to begin destroying the U.S. chemical agent and munitions stockpile. In 1987, the Committee on Review and Evaluation of the Army Chemical Stockpile Disposal Program (Stockpile Committee) was formed. Since that time, the committee has monitored the progress of the Army's Chemical Stockpile Disposal Program (CSDP). Throughout the development of the CSDP, the Stockpile Committee has provided oversight, review, and comment on relevant issues, including the engineering, verification (or systemization), and operations, at both a prototype facility at Johnston Atoll, in the Pacific Ocean, and the Tooele Chemical Agent Disposal Facility (TOCDF), in Utah, the first full-scale chemical agent disposal facility in the continental United States.

Minimizing the risk to workers, the public, and the environment from the continued existence of the stockpile and selecting safe and efficient means of disposal have been, and continue to be, the central themes of the Stockpile Committee's oversight role. Any attempt to minimize risk implies having confidence in the process used to assign values to various sources of risk (risk factors) and the methods used to analyze, compare, and use these factors in decision making. A comprehensive understanding of the full spectrum of risks is fundamental to sound risk management practices. With this in mind, the Stockpile Committee produced a letter report in 1993 calling on the Army to develop site-specific risk assessments as a way of refining the methodology and results of an earlier probabilistic risk assessment that supported the Army's Final Programmatic Environmental Impact Statement issued in 1988.

As the CSDP has progressed, the Stockpile Committee has advised the Army on the need for up-to-date, state-of-the-art, site-specific risk assessments. The recent focus of this advice has centered on the Deseret Chemical Depot (formerly Tooele Army Depot, South), where about 45 percent of the U.S. chemical agent and munitions stockpile is stored, and on the Tooele Chemical Agent Disposal Facility, the associated disposal facility. The remainder of the stockpile is distributed among seven continental U.S. storage sites and Johnston Atoll.

The consensus of the Stockpile Committee is that a clear picture of various risk assessment and risk management activities for Deseret Chemical Depot and the Tooele Chemical Agent Disposal Facility (DCD/TOCDF) has emerged and warrants comment on both the quality of the technical risk analyses and on the Army's integration of these assessments into a comprehensive risk management plan for the site.

The risk management plan is the Army's framework for including considerations of risk in both site-specific and programmatic decisions. The committee's report is intended to facilitate understanding and to encourage public dialogue concerning the DCD/TOCDF and the Army's broader risk management program. In this report, the Stockpile Committee analyzes the chosen risk quantification methodologies and the plan by which the resultant risks are to be managed at the depot and in the disposal facility. Suggestions for improving the risk assessment/risk management process that may be applicable to other chemical storage sites and to the overall disposal program are also made.

The committee greatly appreciates the support and assistance of National Research Council staff members Donald L. Siebenaler, Shirel R. Smith, and Carol R. Arenberg, as well as NRC consultant Harrison T. Pannella, in the production of this report.

RICHARD S. MAGEE, *chair*
ELISABETH M. DRAKE, *vice chair*
Committee on Review and Evaluation of the
Army Chemical Stockpile Disposal Program

Contents

EXECUTIVE SUMMARY 1

1 INTRODUCTION AND BACKGROUND 8
    Description of the Chemical Agent and Munitions Stockpile, 8
    Call for Disposal, 8
      Chemical Stockpile Disposal Program, 8
      Chemical Weapons Convention, 9
      Selection and Development of the Baseline Incineration System, 10
      Historical Risk Assessment by the Chemical Stockpile Disposal Program, 10
    Role of the National Research Council, 10
      Committee on Review and Evaluation of the Army
      Chemical Stockpile Disposal Program, 10
      Composition of the Stockpile Committee, 12
    Purpose of the Report, 12

2 DESERET CHEMICAL DEPOT/TOOELE CHEMICAL AGENT DISPOSAL
   FACILITY SITE-SPECIFIC RISK ASSESSMENTS 14
    Overview, 14
      Deseret Chemical Depot Stockpile, 14
      Sources of Risk, 14
      Risk Receptors, 16
      Risk Measures, 16
      Risk Mitigation, 16
    Objectives and Scope of the DCD/TOCDF Risk Assessments, 16
      Quantitative Risk Assessment, 17
      Health Risk Assessment, 18
    Approach and Methodology, 18
      Quantitative Risk Assessment, 18
      Health Risk Assessment, 22
    Stockpile Committee Oversight, 22
      Quantitative Risk Assessment, 23
      Health Risk Assessment, 23

Additional Review of the Risk Assessments, 23
  Quantitative Risk Assessment, 23
  Health Risk Assessment, 25
Results, 25
  Quantitative Risk Assessment, 25
  Health Risk Assessment, 33
Keeping Assessments Current, 35
  Quantitative Risk Assessment, 35
  Health Risk Assessment, 36
Analyzing and Integrating Results, 36

3 RISK MANAGEMENT IN THE CHEMICAL STOCKPILE
DISPOSAL PROGRAM 39
  Requirements for Risk Management at DCD/TOCDF, 39
  Evolution of the Risk Management Program, 40
    Current Status, 40
    Further Development of the Draft Risk Management Policy Guide, 44
  Applying Risk Assessment Results to Risk Management, 44
    Example 1: The TOCDF Established Configuration, 44
    Example 2: Carbon Filter System for the Pollution Abatement System, 48

4 FINDINGS AND RECOMMENDATIONS 50
  Overview, 50
  Findings, 51
    Risk Assessments, 51
    Risk Management, 51
    Application of Change Policy to the PAS Carbon Filters, 52
  Recommendations, 53
    Risk Assessments, 53
    Risk Management, 53

REFERENCES 54

APPENDICES

A Perspectives on Risk, Risk Assessment, and Risk Management 59
B Risk Assessment Expert Panel on the Tooele Chemical Agent Disposal
   Facility Quantitative Risk Assessment 75
C Reports of the Committee on Review and Evaluation of the Army
   Chemical Stockpile Disposal Program (Stockpile Committee) 78
D Biographical Sketches of Committee Members 79

List of Figures, Tables, and Boxes

FIGURES

ES-1   Schematic illustration of TOCDF risk elements, 2
ES-2   Contributors to the average public fatality risk from continued storage at DCD, 3
ES-3   Contributors to the average public fatality risk from processing at DCD and TOCDF, 3
ES-4   Comparison of public risks during processing at DCD and TOCDF, 4
ES-5   Contributors to the average risk of fatality to disposal-related workers at DCD and TOCDF, 4
1-1   Location and size (percentage of remaining stockpile) of eight continental U.S. storage sites, 9
2-1   Schematic illustration of risk elements at the TOCDF, 15
2-2   Overview of QRA process, 19
2-3   Rocket handling system fault trees for agent spilled during shear operation, 20
2-4   Contributors to the average public fatality risk from continued storage at DCD, 25
2-5   Public acute fatality risk of DCD stockpile storage over 7.1 years of disposal processing, 26
2-6   Radial polar grid of surrounding population, 27
2-7   Mean public acute fatality complementary cumulative distribution function for munition storage during the 7.1 years of disposal processing, by distance from DCD, 28
2-8   Comparison of public risks during processing at DCD and TOCDF, 29
2-9   Comparison of public risks during processing at DCD and TOCDF (logarithmic scale), 30
2-10   Contributors to the average public fatality risk from processing at DCD and TOCDF, 30
2-11   Contributors to the average risk of fatality to disposal-related workers at DCD and TOCDF, 31
2-12   Summary of mean public risk from storage and processing at DCD and TOCDF, 31
2-13   Public societal acute fatalities for all campaigns (TOCDF disposal processing), 32
2-14   Mean public acute fatality risk by distance from TOCDF during disposal processing, 33
2-15   Acute fatalities for other on-site workers at TOCDF from accidents during disposal processing, 34
3-1   PMCD's organizational elements directly related to risk management, 42
3-2   The change process, 43
A-1   Form of the results: scenario probability, 64
A-2   Aleatory and epistemic uncertainty, 64
A-3   Risk profiles with the same expected risk, 67
A-4   Risk curve, 68
A-5   Form of the results: risk profiles, 69

TABLES

2-1 Summary of the Human Health Risk—Overall Risk of Cancer for Combined TOCDF and CAMDS Disposal Operations, 35
2-2 Risks for an Individual Living 2 to 5 Kilometers from the TOCDF, 37
2-3 Expected Number of Fatalities (Societal Risk), 37
3-1 Issues and Factors in Assessing the Value of Change Options, 44
3-2 Activities by Risk Management Function, 45
3-3 PMCD Risk Management through Its Organizations and Functions, 45
A-1 Scenario List with Cumulative Probability, 68

BOX

2-1 Individual Risk at DCD and the TOCDF in Perspective, 36

Acronyms

APET	accident progression event tree	LPG	liquid propane gas
CAC	Citizens Advisory Commission	MPF	metal parts furnace
CAMDS	Chemical Agent Munitions Disposal System	NRC	National Research Council
CSDP	Chemical Stockpile Disposal Program		
CSEPP	Chemical Stockpile Emergency Preparedness Program	OSHA	Occupational Safety and Health Administration
CWC	Chemical Weapons Convention	OVT	operational verification testing
DCD	Deseret Chemical Depot	PAS	pollution abatement system
DSHW	Division of Solid and Hazardous Waste (Utah)	PFS	PAS filter system
		PMCD	Program Manager for Chemical Demilitarization
EG&G	Edgerton, Germerhausen and Grier, Incorporated	POD	process operational diagram
EPA	Environmental Protection Agency	QRA	quantitative risk assessment
FPEIS	Final Programmatic Environmental Impact Statement	RCRA	Resource Conservation and Recovery Act
GA	nerve agent (Tabun)	SAIC	Science Applications International Corporation
GB	nerve agent (Sarin)		
H, HD, HT	blister or mustard agents	TOCDF	Tooele Chemical Agent Disposal Facility
HRA	health risk assessment		
JACADS	Johnston Atoll Chemical Agent Disposal System	VX	nerve agent

Executive Summary

OVERVIEW

Chemical Stockpile Disposal Program

Public Law 99-145, enacted by Congress in 1985, authorized the Army to initiate the process of eliminating the aging U.S. chemical weapons stockpile and led to the establishment, in 1987, of the National Research Council (NRC) Committee on Review and Evaluation of the Army Chemical Stockpile Disposal Program (Stockpile Committee) to provide the Army with technical advice and program oversight. Subsequently, Public Law 104-484 directed the Army to dispose of the entire unitary chemical agent and munitions stockpile by December 31, 2004.

The Army selected incineration as the baseline destruction technology for the Chemical Stockpile Disposal Program (CSDP), and in 1990, a prototype destruction facility was completed at Johnston Island in the Pacific Ocean. Incineration had been endorsed as the preferred technology in a 1984 report by a predecessor of the Stockpile Committee, the Committee on Demilitarizing Chemical Munitions and Agents. At that time, incineration was selected from several alternatives as the most mature and proven technology for the destruction of agents and munitions. Concurrent with the construction of the Johnston Atoll Chemical Agent Disposal System (JACADS), the Army developed and issued its Final Programmatic Environmental Impact Statement (FPEIS), which included a programmatic risk assessment.

A comprehensive understanding of the full spectrum of risks is fundamental to sound risk management. Risks to the public, workers, and the environment have been and continue to be a central theme of NRC oversight. The Stockpile Committee reviewed the FPEIS and, in a January 1993 letter report, noted that the programmatic risk assessment "was not directed at managing risk at any specific site." In that same report, the committee recommended that the CSDP should include site-specific risk assessments for each of the eight continental U.S. disposal sites as a basis for overall risk management of the disposal program. The report laid out a specification for site-specific risk assessments. Subsequently, in a 1994 report, the Stockpile Committee expanded on the nature of and need for site-specific assessments.

In response to the 1993 letter report, the Army Program Manager for Chemical Demilitarization (PMCD) directed that a quantitative risk assessment (QRA) and a risk management program be developed for each of the continental U.S. storage and disposal sites, beginning with the Deseret Chemical Depot (DCD) (formerly Tooele Army Depot, South) in Tooele, Utah, and its associated Tooele Chemical Agent Disposal Facility (TOCDF).[1] The TOCDF, the first full-scale chemical agent and munitions disposal facility in the continental United States, is a second-generation incineration system that incorporates design modifications to improve performance over the initial full-scale prototype incineration system used at JACADS. The stockpile at DCD contains neurotoxic (nerve) agents (GB, VX, GA) and mustard (blister) agents (H, HD, HT) in bulk (ton) containers, rockets, projectiles, mines, bombs, cartridges,

---

[1] U.S. Army. 1996c. Tooele Chemical Agent Disposal Facility Quantitative Risk Assessment. SAIC-96/2600. Aberdeen Proving Ground, Md.: U.S. Army Program Manager for Chemical Demilitarization.

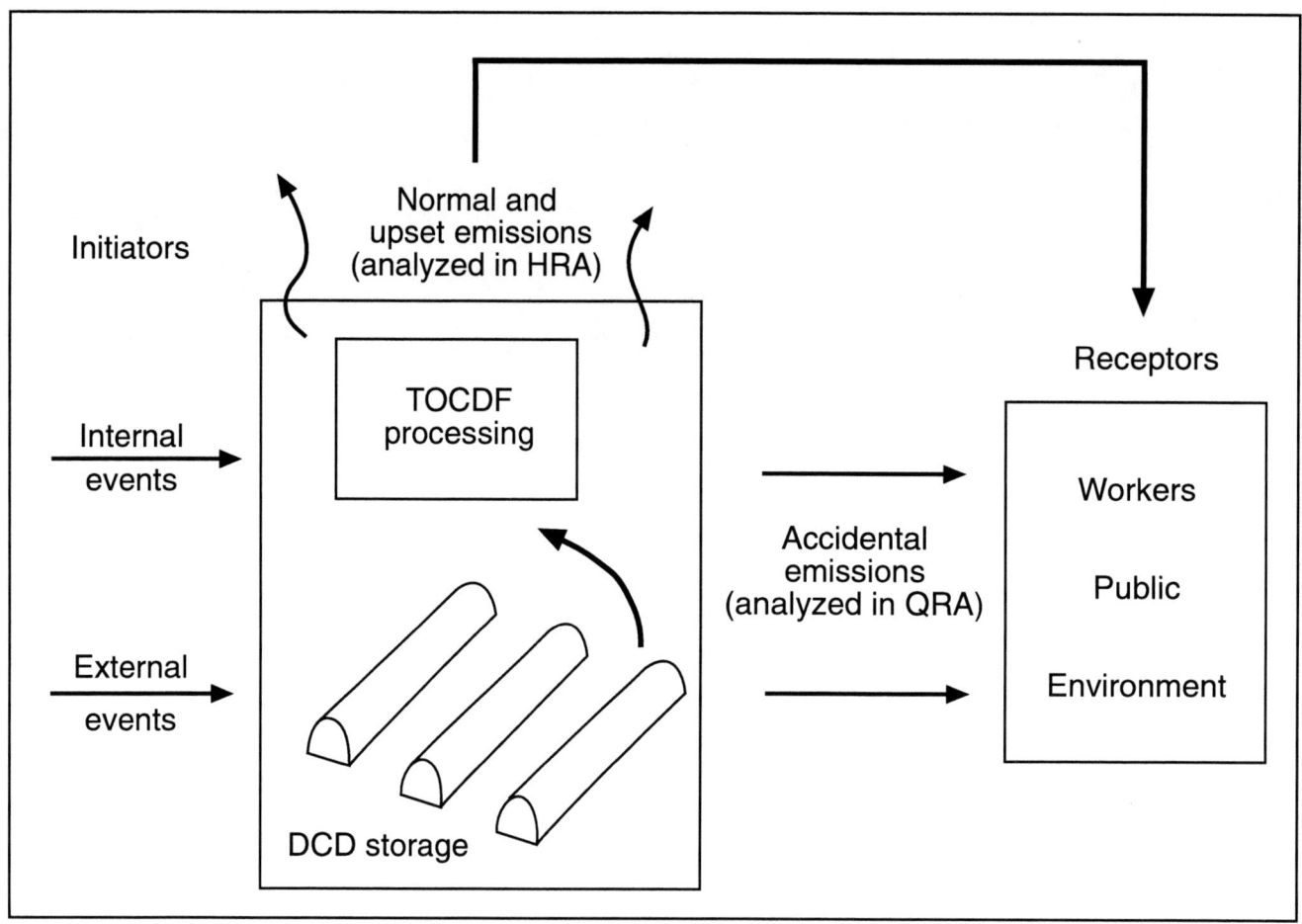

FIGURE ES-1 Schematic illustration of TOCDF risk elements.

and spray tanks. The DCD contains 45 percent of the remaining U.S. stockpile, some 13,000 tons of agent in more than one million inventory items. Agent GB, the most hazardous agent stored at DCD, constitutes about half of the agent and is contained in 75 percent of the inventory items.

A health risk assessment (HRA) for TOCDF was conducted by the state of Utah as part of the environmental permitting process.[2] Figure ES-1 is a schematic illustration of the two major categories of risks, i.e., risks arising from accidents (analyzed in the QRA) and risks arising from emissions during normal and mild upset conditions (analyzed in the HRA). These first site-specific risk assessments have been completed and are primary information for this report.

The Army's site-specific QRA for DCD/TOCDF was conducted, under contract, by Science Applications International Corporation, Inc. (SAIC). An independent group of experts in risk assessment and engineering, the Risk Assessment Expert Panel on the Tooele Chemical Agent Disposal Facility Quantitative Risk Assessment (Expert Panel), oversaw the QRA and provided the Army with a separate, detailed technical review. Thus, the committee's role included both general oversight of the risk assessment/risk management process and oversight of the Expert Panel review process. One or two members of the committee attended most Expert Panel review meetings. The committee also took advantage of many opportunities to examine the technical details of the risk assessment work. A 1996 Stockpile Committee report reviewed the

---

[2]Utah DSHW (Division of Solid and Hazardous Waste). 1996. Tooele Chemical Demilitarization Facility Screening Risk Assessment. EPA I.D. No. UT5210090002. Salt Lake City, Utah: Department of Environmental Quality.

methodology of the QRA and the Expert Panel's oversight. The committee found that:

- The QRA methods met the recommendations of the committee's earlier reports.
- The SAIC QRA team was responsive to the committee's questions and the Expert Panel's comments. New analytical tools for first-of-a-kind QRA calculations were developed; outside experts were retained to give advice in areas where the literature was incomplete; tests and new mechanistic analyses to answer technical questions were conducted; and the QRA analyses were being revised based on that new information.

The current report continues the Stockpile Committee's oversight of risk considerations. The report encompasses the program-wide and site-specific definition of the documented CSDP risk management process and evaluates the results of risk assessments of the Tooele storage and disposal facilities and the overall risk management process being implemented for the TOCDF.

RESULTS OF THE RISK ASSESSMENTS

Quantitative Risk Assessment

Stockpile Storage

Earthquakes as initiating events pose the greatest risk to the public from continued storage of the DCD stockpile. Earthquakes can have widespread, often severe effects, leading to significant adverse consequences. Seismic events that contribute to public risk have mean accelerations higher than 0.2 g (1 g equals the acceleration of gravity) and recurrence intervals of 1,000 years or more. Such earthquakes significantly exceed normal building code design values and thus can lead to failures of equipment, structures, and stored munitions. Earthquakes account for 82 percent of the average risk of fatalities to the public. Significant contributors to the public risk from storage are illustrated in Figure ES-2.

Operational Risk

Public risk from disposal processing during the 7.1-year disposal period is substantially lower than the risk

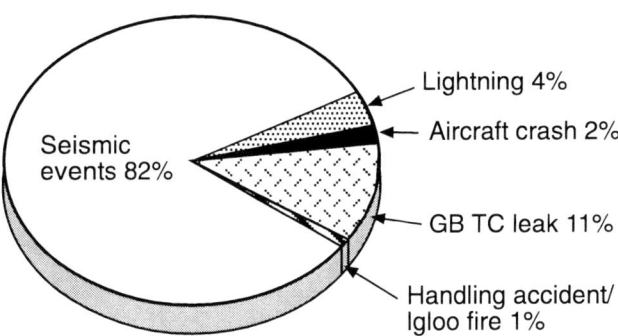

FIGURE ES-2 Contributors to the average public fatality risk from continued storage at DCD.

from continued storage during the same period. Again, earthquakes are the most significant initiating events, accounting for 97.4 percent of the public risk during processing. Other contributors to risk are summarized in Figure ES-3. Figure ES-4 depicts the magnitude of risk associated with continued storage and agent/munitions processing and shows the decrease in storage risk as the stockpile is processed. This figure provides the most complete picture of the risks during processing and the most thorough comparison of the risks of continued storage with the risks associated with processing (including the diminishing contribution of storage risk during the processing period).

For the 500 workers at the TOCDF, there is about a one in seven probability that there will be one fatality

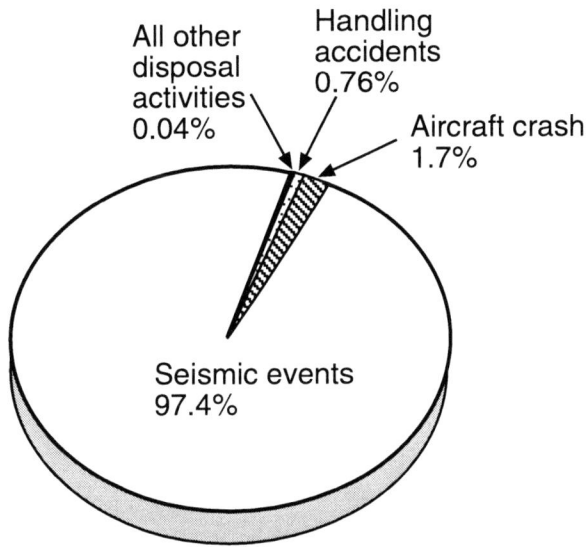

FIGURE ES-3 Contributors to the average public fatality risk from processing at DCD and TOCDF.

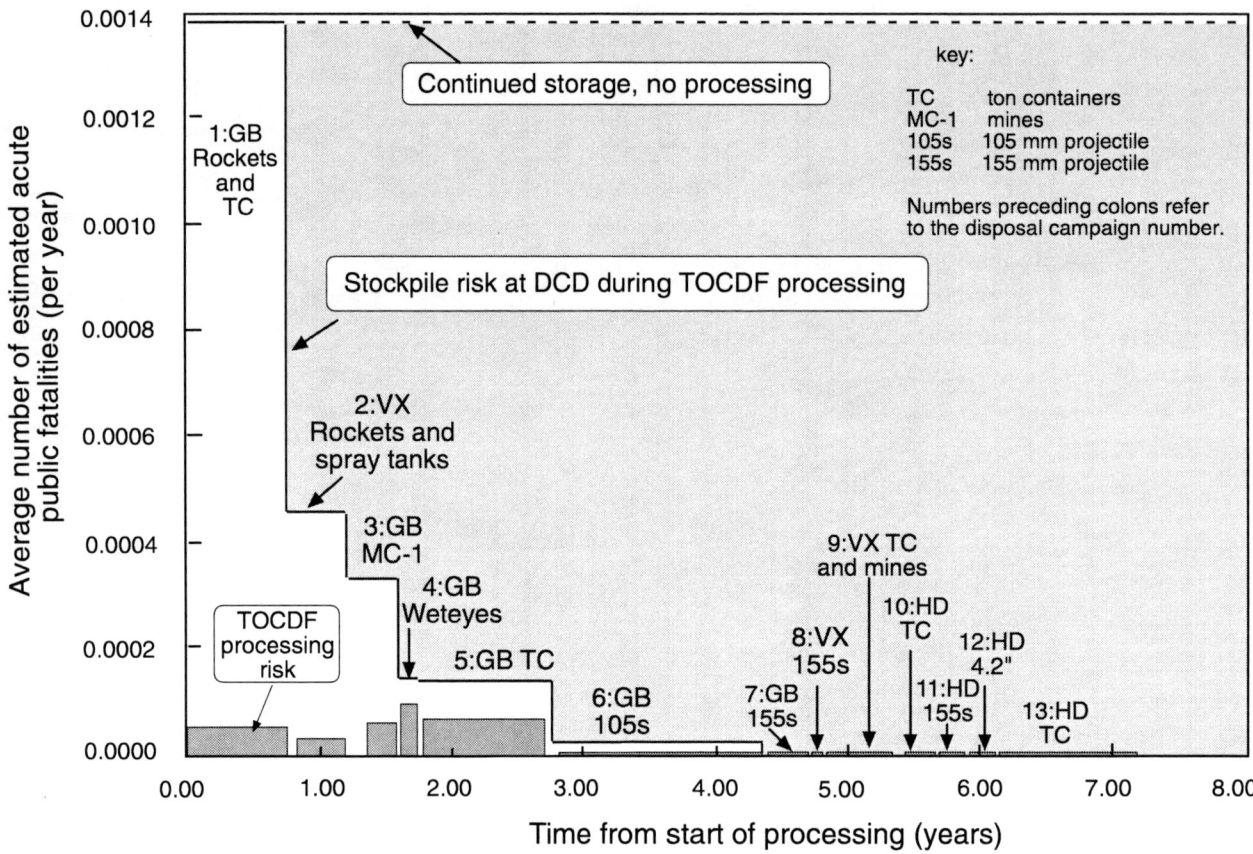

FIGURE ES-4 Comparison of public risks during processing at DCD and TOCDF.

in the 7.1 years of disposal processing (expected number of fatalities is 0.13). This is equivalent to the probability of fatality to an individual worker of one in 25,000 (or 4×10^{-5}) per year, which is consistent with the average occupational risk for all occupations in the United States. However, because the risk levels to disposal workers from agent exposure must be added to normal occupational risk levels, the committee believes that emphasis on job safety related to both agent and nonagent activities is extremely important. The sources of risk for workers are shown in Figure ES-5.

Health Risk Assessment

The HRA evaluated the effects of incinerator emissions under various operating scenarios for an adult resident, a child resident, a subsistence fisherman, and three farmers at various locations with various durations of exposure during the year. Conservative modeling assumptions maximizing anticipated concentrations of emissions and maximizing exposure values at points of maximum off-site impact were used to derive the upper-end estimates of risk. The calculated human health risk was measured against the 1×10^{-5} carcinogenic risk level established under EPA exposure assessment guidance protocols. This threshold was not exceeded in any scenario.

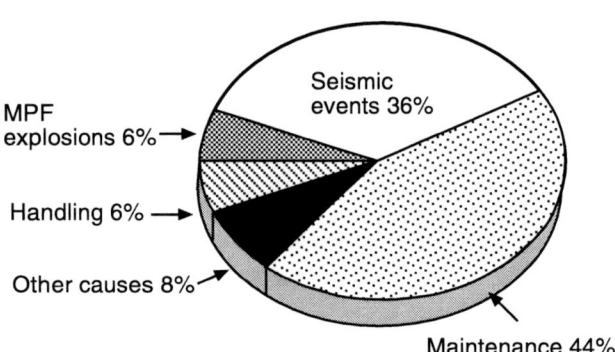

FIGURE ES-5 Contributors to the average risk of fatality to disposal-related workers at DCD and TOCDF.

RISK MANAGEMENT IN THE CHEMICAL STOCKPILE DISPOSAL PROGRAM

Risk management typically involves the following steps: understanding the risk; suggesting ways to reduce risk; evaluating the alternatives; and selecting preferred alternatives. Each affected party plays a role in the risk management process. The Army is responsible for managing the chemical stockpile and its destruction in a manner that maintains safety for the public, workers, and the environment. The current CSDP risk management program is a multilevel program that delineates policy, defines requirements, provides guidance for implementation, defines specific requirements for the facility, and defines the processes that should be used.

The Army recently published (in draft) *A Guide to Risk Management Policy and Activities* (the *Guide*), which includes a description of the "management of change process," a significant aspect of the risk management process at the TOCDF and other sites. The management of change process may involve changes or modifications to equipment, operating and maintenance procedures, or agent destruction schedules.

The first full scale application of the management of change process is expected to be an evaluation of adding a carbon filter system to the pollution abatement system (PAS). Consideration of the PAS filter system (PFS) was prompted by a committee finding in a 1994 report that adding a carbon filter system downstream of the existing PAS would add further protection against emissions of agent and trace organics, even in the unlikely event of a substantial system upset.

FINDINGS

The Stockpile Committee has followed the DCD/TOCDF QRA project closely since its inception and has maintained oversight of the Expert Panel independent peer review process. The QRA has achieved the goals set out in the committee's 1993 letter report and subsequent reports. The findings in the DCD/TOCDF QRA are consistent with interim findings in the committee's review of systemization of the TOCDF. The committee concurs with the following findings of the Expert Panel:

- The methodology was sound and has extended the state of the art in several areas.
- The methodology was well implemented.
- Despite some reservations concerning a few technical aspects of this QRA, the panel was reasonably satisfied that these did not affect the overall conclusions of the QRA.

The committee finds that the interactive independent review process was effective and that the Expert Panel played a significant role in ensuring that the QRA met or exceeded state-of-the-art standards in all significant respects.

The committee believes that the HRA performed by the Utah Division of Solid and Hazardous Waste, which is based on many assumptions and follows EPA-mandated protocols, is appropriate at this stage of TOCDF operations because it approximates a worst case for the public for all evaluated parameters. The greatest uncertainty in the HRA is about the magnitude and composition of actual TOCDF emissions (emissions in the HRA were based on adjusted data from JACADS). As actual TOCDF operating parameters are established and data on the nature and magnitude of actual emissions become available, they can be incorporated into the HRA. The HRA does not provide the more realistic and detailed estimation of risk sources, impacts, and distribution provided by the QRA. However, it does screen latent cancer risk to "maximally exposed" individuals, impose an acceptability criterion (1×10^{-5} carcinogenic risk level over a 70-year lifetime), and infer that the exposure of multiple individuals at or below the screening level is acceptable.

Risk Management

The committee finds that the TOCDF risk management plan has progressed and that positive steps have been taken, e.g., the development and limited use of guidance and implementation documents. The Army's draft *Guide* on risk management provides an overview of the overall risk management program, incorporating references to subsidiary risk management documents and activities. The *Guide* defines interrelationships among Army offices, contractor offices, and public entities that are or should be involved in risk management activities. The *Guide* is a significant step by the

Army toward following NRC recommendations on risk management and public involvement, particularly with respect to using risk analysis in the management of change process. The *Guide*, however, has not yet been completed and finalized. The committee's comments in this report reflect the expectation that the *Guide* will be improved, adopted, and implemented. The committee may evaluate its implementation in the future, but at present, it falls short of expectations in several respects:

- The *Guide* does not describe the contributions of risk assessment to changes that led to the current Established Configuration, against which proposed future changes will be evaluated.
- The focus of the *Guide* is primarily on agent-related safety rather than on developing and institutionalizing a comprehensive safety program (i.e., establishing a safety culture).
- The *Guide* acknowledges that more work must be done to shift the focus from public information to public involvement. The role of public involvement should be extended and integrated beyond the management of change process.
- The *Guide* is not specific enough about how to ensure that workers, Chemical Stockpile Emergency Preparedness Program (CSEPP) personnel, and the public understand the risk analyses.
- Although the *Guide* indicates that there are functional relationships among PMCD organizations with regard to risk management, it does not detail how management roles and communications across all the Army groups, contractors, subcontractors, and other agencies involved in the program will be integrated.
- The *Guide* does not indicate plans to track CSEPP responses to changes in risk or to document public involvement and Army responses to public input regarding risk.

On the positive side, the *Guide* presents a framework for managing changes in the configuration of a facility or changes in operations that may significantly affect risk levels. The framework allows for public input on significant changes through a comment process, which is followed by formal feedback to the public explaining the basis for a decision. The committee finds the proposed management of change process satisfactory and encourages its use. However, there may be opportunities to further expand public participation as the Army develops a more comprehensive public involvement program.

Application of Change Policy to PAS Carbon Filters

Concerning the evaluation of adding carbon filters to the PAS (an earlier recommendation of the Stockpile Committee), the committee finds that:

- The proposed methodology, if well implemented, is appropriate for evaluating whether or not to install a PFS on a site-specific basis.
- The proposed methodology for the PFS evaluation is consistent with the Army's proposed management of change process, as described in the *Guide*.
- Carbon filters appear to be effective in reducing the levels of dioxins/furans to below the limits of detection and have a useful life of at least one year. Because these levels are too low to be measured, however, credit only for reducing them to the detection limit appears in the HRA.
- The QRA calculations for the PFS must account for a potential sudden release of accumulated agent (based on HRA-assumed emissions at the lower detection limit) in case of a PFS malfunction.

RECOMMENDATIONS

Risk Assessments

Recommendation 1. The Army should update both the QRA and HRA at the TOCDF whenever changes to system design or operations occur that could affect QRA or HRA calculations to ensure that estimates of risk are current and reflect changes in operating conditions and experience, assumptions, and program status (current Established Configuration). The process for updating the QRA and HRA should be included in the *Guide*.

Recommendation 2. The Army should continue the site-specific QRA and HRA processes at all PMCD sites. The development of assessments for sites other than the DCD will be greatly simplified because much of the methodology has already been established. The Army should continue to obtain interactive, independent expert reviews of all site-specific risk assessments. The Army should heed the lessons learned from development of the TOCDF QRA and should incorporate the changes recommended by the Expert Panel.

Recommendation 3. The QRA methodology manual should be updated to reflect the significant improvements that have been made.

Risk Management

Policy

Recommendation 4. The Army should expand its draft report on risk management policy, *A Guide to Risk Management Policy and Activities*, to encourage the establishment of a "safety culture" within the PMCD and its field offices and among contractors and other government agencies. The *Guide* should elucidate the Army's policy on industrial safety, including the responsibilities of individuals and managers in the field and the definitions of acceptable performance.

Recommendation 5. The Army should develop a management program (and include it in the *Guide*) that defines the integration of management roles, responsibilities, and communications across activities by risk management functions (e.g., operations, safety, environmental protection, emergency preparedness, and public outreach).

Recommendation 6. The Army should review and expand the current draft risk management plan to include public involvement in appropriate areas beyond the management of change process.

Recommendation 7. The Army should institutionalize the management of change process developed in the *Guide*. The Army should track performance of the change and document public involvement and public responses to decisions. The Army should use this experience to improve the change process.

Recommendation 8. The Army should expand implementation of the risk management program to ensure that workers understand the results of the risk assessments and risk management decisions. The Army should also ensure that CSEPP and other emergency preparedness officials understand the QRA and how their activities might affect risk. CSEPP activities should be tracked by the Army as part of the risk management program.

Recommendation 9. The Army should implement the risk management plans and update them whenever necessary to ensure that they reflect current practices and lessons learned.

Evaluation of the Carbon Filter Design for the Pollution Abatement System

Recommendation 10. The Army should proceed with the application of its proposed methodology for evaluating the use of PAS carbon filters on a site-specific basis. For consistency with the HRA assumptions, the QRA should take into account the possible sudden release of agent that may have accumulated on the filter at a gas concentration equal to the lower detection limit.

1

Introduction and Background

DESCRIPTION OF THE CHEMICAL AGENT AND MUNITIONS STOCKPILE

For more than 50 years, the United States has maintained a stockpile of chemical agents and munitions distributed among eight sites within the continental United States and at Johnston Atoll in the Pacific Ocean (Figure 1-1). Two basic types of chemical agents comprise the stockpile: neurotoxic (nerve) agents and mustard (blister) agents. Both types are frequently, and erroneously, referred to as "gases" even though they are liquids at normal temperature and pressure.

The nerve agents include the organic phosphorus compounds designated as VX, GB (Sarin), and GA (Tabun). These chemicals present a significant toxic hazard because of their action on the nervous systems of humans and animals through inhibition of the acetylcholinesterase enzyme. They are both considered extremely toxic. VX is more acutely toxic than GB, but the latter represents a greater potential hazard because of its higher volatility (about the same as water) and, thus, the greater likelihood of its being inhaled. Chronic health effects and cancer from low-level exposures have not been associated with nerve agents or with chemically (and toxicologically) similar commercially available organic phosphorus insecticides (Leffingwell, 1993). Only short-term symptoms have been documented in individuals who survive exposure to nerve agents.

The mustards (designated H [nondistilled mustard], HD [distilled mustard], and HT [thickened mustard]) do not present significant acute lethal hazards. Their principal effect is severe blistering of the skin and mucous membranes. They have been implicated as being carcinogenic, however, and may present a cancer hazard to individuals exposed acutely (Leffingwell, 1993; IOM, 1993). The estimates for induced cancers from accidental agent exposures (Chapter 2) only consider mustard agents.

Chemical agents, after being fully dispersed, do not tend to persist in the environment because their relatively simple chemical structures tend to undergo hydrolysis in humid climates. However, in extremely dry desert climates, they can remain for a considerable period of time (U.S. Army, 1988).

The chemical agents in the U.S. stockpile are stored in a variety of containment systems, including bulk (ton) containers, rockets, projectiles, mines, bombs, cartridges, and spray tanks. Figure 1-1 summarizes the stockpile configuration as of 1996 for the eight continental U.S. sites by agent, munition, and containment system (OTA, 1992; NRC, 1996a).

CALL FOR DISPOSAL

Chemical Stockpile Disposal Program

Because of the age of the chemical weapons stockpile, their lack of utility as weapons or deterrents, the continuing costs of maintenance, and the potential for accidental release, there is now sufficient incentive for the United States (and other countries) to dispose of stored chemical weapons. In 1985, Congress enacted Public Law 99-145 to initiate the process of eliminating the U.S. chemical weapons stockpile, with an expedited program to dispose of M55 rockets, which raise

INTRODUCTION AND BACKGROUND

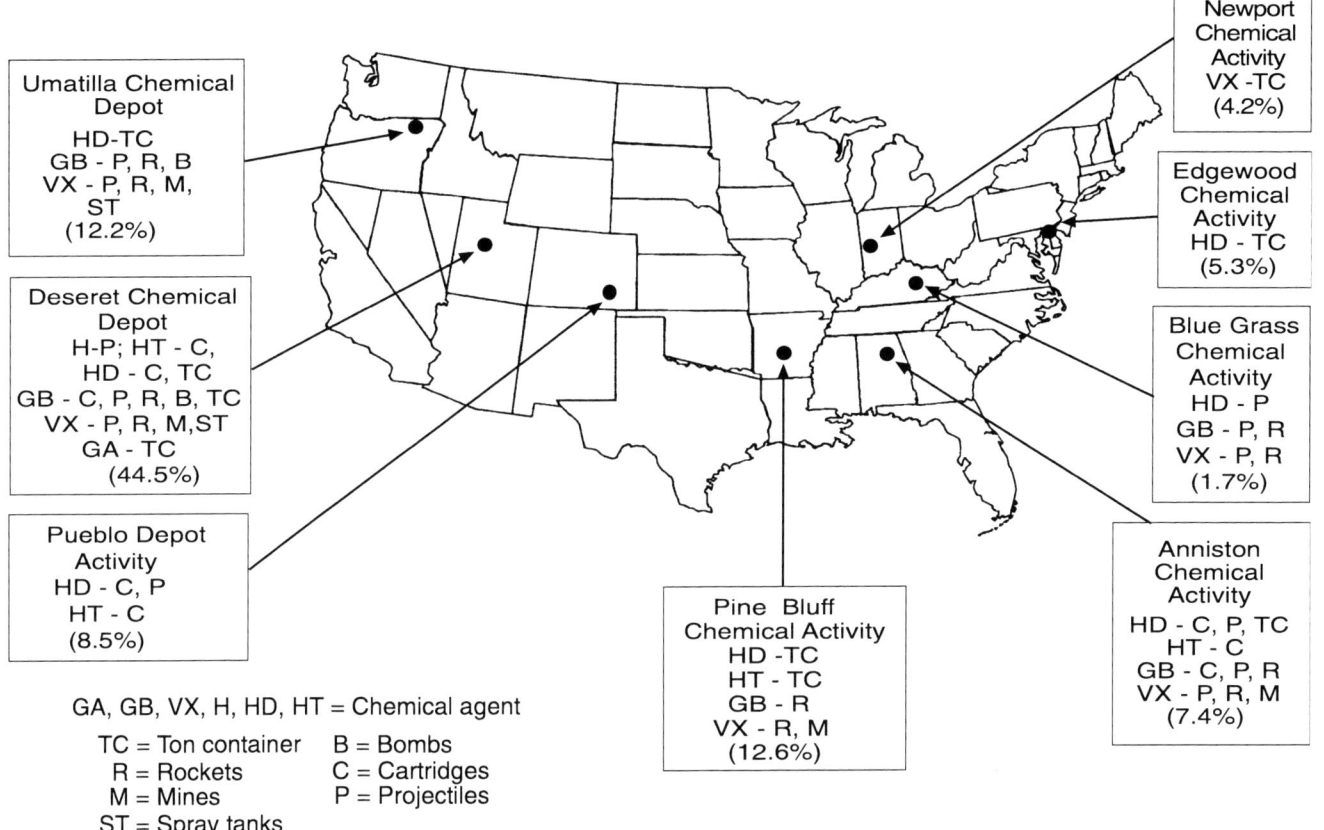

FIGURE 1-1 Location and size (percentage of remaining stockpile) of eight continental U.S. storage sites. Source: OTA, 1992; NRC, 1996a.

special concerns because they are aging and because they contain agent, explosives, and propellant in an integrated configuration. Later, in 1992, Congress enacted Public Law 104-484, which directed the Army to dispose of the entire unitary[1] chemical agent and munitions stockpile by December 31, 2004. Congress also directed that the Chemical Stockpile Disposal Program (CSDP) be implemented in a manner that ensured maximum protection of workers, the public, and the environment.

Chemical Weapons Convention

The CSDP has evolved in parallel with worldwide activities addressing questions of international control and the elimination of chemical agents and munitions. Over the course of several decades, a broad and complex agreement known as the Chemical Weapons Convention (CWC) was negotiated. Since 1993, the CWC has been signed by 165 countries and ratified by 89 countries. The convention was to go into effect six months after 65 countries had ratified it, which occurred on October 29, 1996. The CWC entered into force on April 29, 1997. The United States was actively involved in negotiating the CWC agreement and recently ratified it. Russia, the world's largest holder of chemical agents and munitions, has yet to ratify it.

The CWC defines the destruction of chemical weapons as "a process by which chemicals are converted in an essentially irreversible way to a form unsuitable for production of chemical weapons, and which, in an irreversible manner, renders munitions and other devices unusable as such" (Smithson, 1993). The method of destruction is to be determined by each country, but the manner of destruction must ensure public safety and protect the environment.

---

[1]The term *unitary* refers to a single chemical loaded in munitions or stored as a lethal material. More recently binary munitions have been produced, in which two relatively safe chemicals are loaded into separate compartments to be mixed to form a lethal agent after the munition is fired or released. The components of binary munitions are stockpiled separately, in separate states. They are not included in the present Chemical Stockpile Disposal Program. However, under the Chemical Weapons Convention of 1993, they are included in the munitions that will be destroyed.

The CWC prohibits the development, production, acquisition, stockpiling, retention, transfer, or use of chemical weapons. Article IV of the CWC requires that signatories destroy chemical weapons and any special facilities for their manufacture within 10 years, i.e., by April 29, 2007. The date established by Congress for the destruction of the U.S. chemical stockpile remains December 31, 2004.

Selection and Development of the Baseline Incineration System

In the early 1980s, the Army investigated a number of technologies and strategies for the destruction or disposal of chemical weapons. Among these were chemical neutralization, ocean disposal (now banned by federal law), stockpile consolidation with subsequent destruction, and disassembly followed by component incineration. Incineration was selected by the Army as the preferred technology for stockpile disposal. The National Research Council (NRC) Committee on Demilitarizing Chemical Munitions and Agents was formed in August 1983 to review the status of the stockpile and to assess the available disposal technologies. In the committee's final report in 1984, incineration was endorsed as an adequate technology for the safe disposal of chemical warfare agents and munitions (NRC, 1984).

Pursuant to the enactment of Public Law 99-145, the Army began development of components of the baseline incineration system at the Chemical Agent Munitions Disposal System (CAMDS) facility at Deseret Chemical Depot (DCD), formerly Tooele Army Depot South, Utah. Construction and systemization of the first fully integrated baseline incineration system, the Johnston Atoll Chemical Agent Disposal System (JACADS), was completed in July 1990 on Johnston Island, located in the Pacific Ocean approximately 700 miles southwest of Hawaii. The JACADS facility has a twofold mission:

- to destroy the chemical agents and munitions stored there
- to serve as a demonstration facility for the baseline incineration system

Historical Risk Assessment by the Chemical Stockpile Disposal Program

At a relatively early stage of the CSDP, a probabilistic risk assessment was performed in support of the Army's decision to use a baseline incineration system on site (U.S. Army, 1987). The PRA was documented in the Final Programmatic Environmental Impact Statement (FPEIS) (U.S. Army, 1988). The probabilistic risk assessment (PRA) done at that time was a less detailed version of the quantitative risk assessment (QRA).

The FPEIS PRA evaluated accident sequences that could result in agent releases during the disposal process based upon the system design for JACADS. The PRA also examined risks for several disposal/transportation options at the eight continental U.S. storage sites. The JACADS analysis was modified slightly to account for major site-to-site differences. However, it was not site-specific in its treatment of design differences or local operating and maintenance practices, including disposal scheduling. The analyses of site-specific external-event hazards scenarios and the treatment of handling accidents, other particular accidents, and uncertainty were also less thorough.

ROLE OF THE NATIONAL RESEARCH COUNCIL

Committee on Review and Evaluation of the Army Chemical Stockpile Disposal Program

Concurrent with construction at JACADS, in 1987 the Army requested that the NRC review and evaluate the Army CSDP. The NRC established the Committee on Review and Evaluation of the Army Chemical Stockpile Disposal Program (Stockpile Committee) to perform these tasks over time and specifically to monitor operational verification testing (OVT) at JACADS, which began in July 1990 and was completed in March 1993. In July 1993, the NRC issued a preliminary short report (Part I) on OVT (NRC, 1993a) and in April 1994, a final report (Part II) on OVT at JACADS (NRC, 1994a). These reports concluded that the baseline incineration system was an adequate and safe means of disposing of the chemical weapons stockpile. Several subsequent reports have reaffirmed the committee's position.

Construction of the first disposal facility in the

continental United States was begun in 1989 in Utah. This facility, the Tooele Chemical Agent Disposal Facility (TOCDF), is a "second-generation" baseline system, which has incorporated into its design and operating procedures many improvements and technological advances based on JACADS operating experience (NRC, 1996b). The recommendations of the Stockpile Committee have been a factor in the changes and improvements to the facility. Pre-operational testing (systemization) at the TOCDF began in August 1993. During systemization, several modifications were made to systems and procedures at the TOCDF, e.g., a new slag removal system was designed for the liquid incinerator to eliminate the need for frequent shutdowns of the unit for the manual removal of slag; the furnace feed system was reviewed in detail by MITRE Corporation, and changes were made to correct misfeed problems that had occurred at JACADS.

In reviewing the FPEIS, the Stockpile Committee recognized the generic nature of the PRA and noted in a letter report (NRC, 1993b) that "the risk analysis as presented in the FPEIS was not directed at managing risk at any specific site." In the same letter report, the committee noted that

> the continental sites at which lethal chemical agents and munitions will be destroyed all differ substantially from Johnston Island, as well as from one another, with regard to terrain, weather, the density of nearby population, the transportation network, the size and variety of stored agents and munitions, other aspects, and, possibly, destruction technology" (NRC, 1993b).

The committee specifically recommended in the letter report that "a site-specific, full-scope, scenario-based risk assessment should be performed for each continental U.S. facility, starting with the Tooele facility" and that "each site-specific risk assessment should include the case of continued storage without disposal as one scenario." The letter report laid out detailed technical specifications for the site-specific risk assessments (NRC, 1993b). Another NRC report, *Recommendations for the Disposal of Chemical Agents and Munitions* (NRC, 1994b), reiterated this recommendation and emphasized the importance of site-specific risk assessments to sound risk management practices.

In response to the NRC's recommendations, the Program Manager for Chemical Demilitarization (PMCD) directed that a QRA and a risk management program be developed for each continental site, beginning with DCD/TOCDF. Concurrently, the Army retained a five-member panel of experts to provide an independent review of the approach and methodology for the QRA.

In the 1996 report, *Review of Systemization of the Tooele Chemical Agent Disposal Facility*, the Stockpile Committee presented an evaluation of the methodology for the site-specific QRA for DCD/TOCDF (NRC, 1996b). The report described the QRA methodology and indicated the committee's approval of the methods used and satisfaction with the QRA team's response to questions and criticisms. The QRA team vigorously pursued gathering new information (new tests, new mechanistic calculations, and expert knowledge) whenever concerns were raised about aspects of their analysis. The *Systemization* report also cited risk-related recommendations from previous NRC reports and evaluated the Army's response up to that time. The committee was satisfied with the role of the Risk Assessment Expert Panel on the Tooele Chemical Agent Disposal Facility Quantitative Risk Assessment (Expert Panel) and the QRA team's diligent response to each comment from the Expert Panel. The report indicated the committee's satisfaction with the way in which the risk assessment addressed all the relevant recommendations, with three exceptions: (1) the analysis was not yet complete; therefore it would be necessary to ensure that the remaining work continued to meet the committee's recommendations; (2) public involvement in the QRA needed to be improved; and (3) there appeared to be a lack of coordination between the QRA and the health risk assessment (HRA). Therefore, the committee also advocated the preparation of a single risk assessment summary report for each site to present integrated results of the various risk studies being conducted as separate projects under diverse auspices.

The first set of site-specific risk assessments for DCD/TOCDF and associated risk management documents have now been assembled and are the basis for this report. The committee has provided additional comments about public involvement relating to risk assessments in a recent report on community involvement (NRC, 1996c). The Army has recently published phase-one (first results) QRAs for five other sites (U.S.

Army, 1995a, 1996a, 1996b, 1997a, 1997b). Although evaluations of these QRAs are not included in this report, they show that the site-specific risk varies widely from site to site with the nature of the stockpile, the demographics near the site boundary, and the potential for external events.

Composition of the Stockpile Committee

The experience and familiarity from advisory and oversight activities, and from report development by the Stockpile Committee since its inception, provide a sound basis upon which to evaluate the present state of the Army's risk management activities for the Tooele storage and disposal facilities. Over the years, the Stockpile Committee has adjusted the composition of its membership to maintain a balance of disciplines necessary to meet the task at hand. Of the 15 current members, two are long-standing, recognized experts in the field of risk assessment and risk management. Other members of the Stockpile Committee have expertise in risk communications, public involvement, chemical engineering, mechanical engineering, combustion technology, biochemical engineering, chemical process design and control, analytical chemistry, toxicology, emergency response, human systems, and environmental law and sciences.

The Stockpile Committee has prepared 16 NRC reports on various aspects of the overall CSDP, the development of the baseline incineration system, the systemization of the TOCDF, and the importance of public involvement. Appendix C is a list of these reports. The baseline incineration system at the TOCDF has evolved over the past decade through refinement of the prototype facilities at CAMDS and JACADS. The Stockpile Committee has tracked developments at these facilities and at the TOCDF and has commented extensively on progress at all of them through the construction, systemization, OVT, and agent destruction phases.

PURPOSE OF THE REPORT

This report continues the oversight of risk considerations begun in previous Stockpile Committee reports. It encompasses the program-wide and site-specific definition of the documented CSDP risk management process and evaluates the results of risk assessments performed for the Tooele storage and disposal facilities and the overall risk management process being implemented for the TOCDF, the first full-scale chemical agent and munitions disposal facility in the continental United States.

In this evaluation, the committee focuses on five areas: (1) the quality of the risk assessments conducted for DCD/TOCDF; (2) the significance and interpretation of results and conclusions; (3) the integration (or lack thereof) of results from separate risk assessments; (4) the utilization of risk assessments in a comprehensive risk management plan for the TOCDF as well as the CSDP; and (5) the implementation of risk management practices.

Characterizing effective risk management processes is the central theme of this report. The emphasis is on the utility of thorough, high quality, technically sound risk assessments as a basis for risk management practices. This report covers both major risk assessments, namely, the QRA (a quantitative evaluation of risks from accidental releases of agent), and the HRA (health risk assessment, which approximates worst case analyses of stack emissions for normal and upset operations). Other data being gathered by the Army at Tooele could be used to decide if a separate agricultural risk assessment will be needed in the future. As for the QRA and the HRA, this report considers the methodology, scope, and technical quality of the analyses; the treatment of acute and latent risks resulting from accidents; the risks from normal and upset operations; and the use of the results for the purpose of risk management. The report also addresses the integration of results of various assessments and how they can be communicated effectively, both within the CSDP and externally, to involve the local community in decisions pertaining to storage and disposal operations.

The assessments of risk at the Tooele site involve a chemical agent and munitions storage and disposal system that includes sophisticated technology, procedural regimes, and contingency plans. Although the committee intends this commentary to assist the Army with the ongoing implementation of an effective risk management program at the Tooele site, a broader goal is to improve the risk assessment and management process at future sites, at both the programmatic and site-specific levels.

The committee hopes that by reporting on the Army's CSDP risk management program at Tooele, the public may come to a better understanding of the risks, the thoroughness with which these risks have been analyzed, and the ways risk analyses have been used to reduce and manage risk. Consequently, in addition to the U.S. Army, the public is viewed as a primary audience for this report.

Chapter 2 assumes the reader's familiarity with the methodology and terminology of risk analysis and assessment. Readers who are not familiar with the subject may wish to begin with Appendix A, which presents an introduction to the subject, starting with the simple example of a person tripping over a crack in the sidewalk. The example is then expanded to include some of the complications and refinements required in a real risk assessment of complex facilities like DCD/TOCDF. A final section of Appendix A discusses the process of risk management for the CSDP.

2

Deseret Chemical Depot/Tooele Chemical Agent Disposal Facility Site-Specific Risk Assessments

OVERVIEW

In this chapter, a brief overview of the sources of risk at DCD/TOCDF and of the risk assessments performed by Science Applications International Corporation (SAIC) and the state of Utah is presented followed by descriptions of the objectives and scope, approach and methodology, oversight, results and analysis, and integration of results of the assessments. SAIC performed a QRA that examined the risk from agent accidents (U.S. Army, 1996c). The state of Utah performed an HRA that examined the maximum risk from normal and upset operations (Utah DSHW, 1996). (See Appendix A for background information on risk assessment). The SAIC methodology is carefully implemented and consistent with previous recommendations of the committee (NRC, 1993b; NRC, 1994b) except that SAIC was not asked by the Army to present an integrated assessment of the QRA and HRA results. Therefore, under each topic, the QRA and the HRA are discussed sequentially. The risks from both the QRA and HRA are discussed at the end of this chapter. Figure 2-1 illustrates some of the elements of risk discussed in this chapter.

Deseret Chemical Depot Stockpile

As of March 1997, the chemical weapons storage facilities at DCD contained nearly 45 percent of the remaining U.S. chemical weapons stockpile, with more than 13,000 tons of agent. The DCD stockpile contains nerve agents GB, VX, and small quantities of GA, as well as all three types of mustard (blister) agents. In addition, every type of U.S. chemical weapon containment system (mines, rockets, projectiles, bombs, ton containers, and spray tanks) is present in the DCD stockpile, with more than one million individual items in the inventory. The DCD stockpile has the largest quantity of chemical agents and the most complex combination of agent/containment systems.

Sources of Risk

For a chemical agent and munitions storage and destruction site like DCD/TOCDF, there are two primary sources of risk: (1) risk associated with the stockpile itself (stockpile risk) and (2) risk associated with destruction of the stockpile (operational risk). The actual risk from either or both sources depends upon whether risk-initiating events occur. Such events can be either internal or external in nature. *Internal risk-initiating events* are events associated with the storage and routine maintenance of the stockpile and with the operation of the destruction facility. *External risk-initiating events* are events not associated with site operations, such as earthquakes, floods, lightning strikes, and airplane crashes. (Note that there are also external risks to the stockpile from war or sabotage, which are reportedly evaluated and managed by specific government agencies and are not considered in publicly available site-specific risk assessments. The Stockpile Committee has not been involved in or reviewed any of these evaluations.)

Stockpile Risk at DCD

The principal hazards associated with the stockpile at DCD are from the inherent toxicity of the anti-

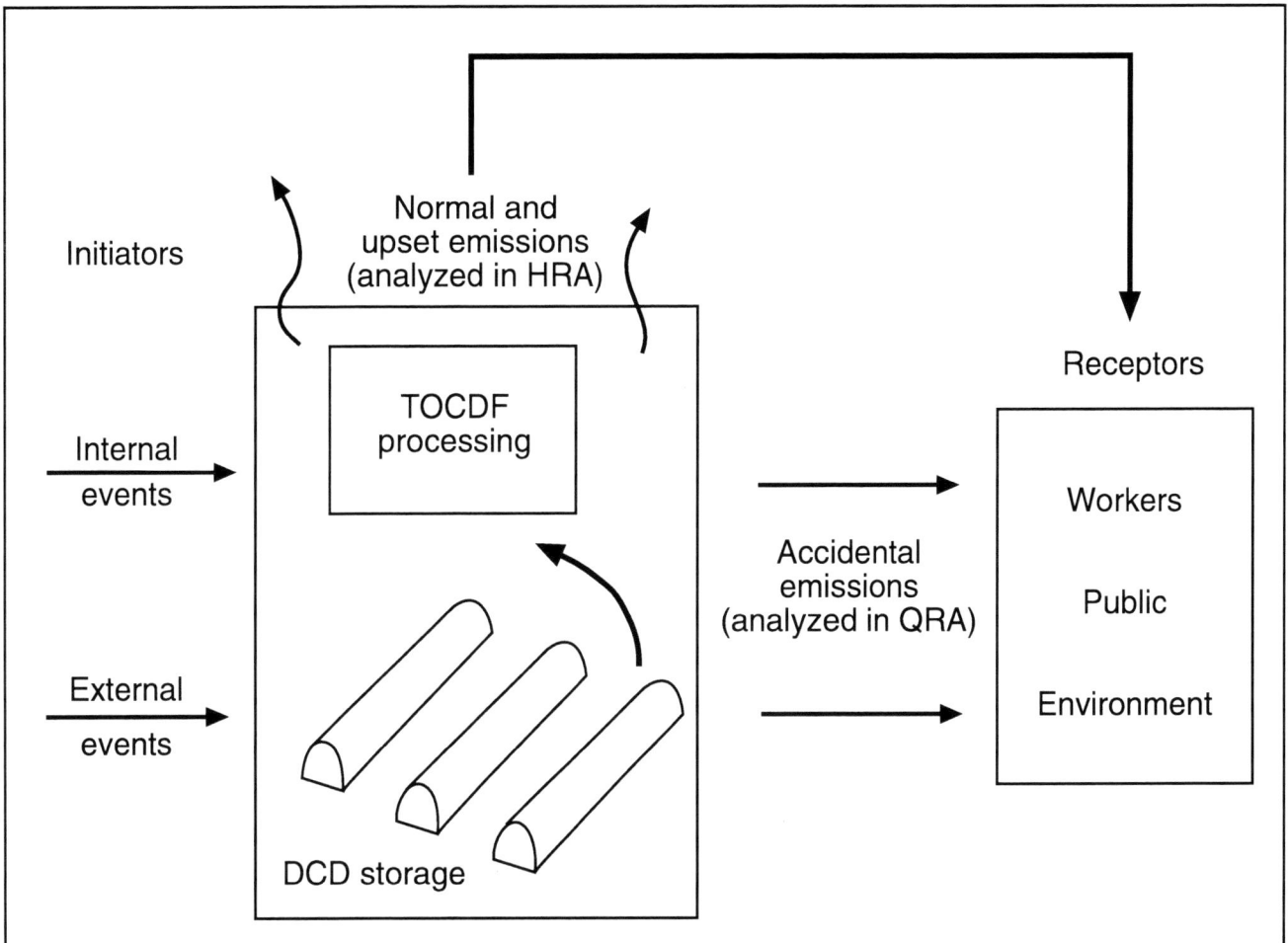

FIGURE 2-1 Schematic illustration of risk elements at the TOCDF.

cholinesterase agents, GB and VX, and mustard agents, H, HD, and HT. Risks associated with the stockpile are almost all related to agent releases from either internal or external events. Agent GB presents the greatest hazard because of its toxicity and volatility; GB also represents the greatest potential risk because it constitutes about half of the total amount of agent on site and is contained in more than 75 percent of the inventory items at DCD.

Agent releases initiated by internal events could result from handling accidents during stockpile manipulation and maintenance; the deterioration of containment systems; the spontaneous detonation of munitions; or the spontaneous ignition of propellant. External events that could cause releases include earthquakes, floods, lightning strikes, and airplane crashes.

Operational Risk at the TOCDF

Agent destruction imposes risks above and beyond the inherent risks associated with the existence and maintenance of the chemical agent and munitions stockpile. The transportation of agents from storage to the destruction facility, the unpacking and disassembly of munitions and containment systems, and the actual agent destruction processes provide additional opportunities for agent releases caused by internal or external events. Like the stockpile risk, the predominant operational risks are associated with agent toxicity. However, the quantities of agent being processed at any given time are small compared to the original inventories in the stockpile. The maximum quantity of agent present in the disposal facility at any given time would be the equivalent of about three ton containers (i.e., approximately 5,000 pounds of agent).

In addition to agent, other risks from the agent destruction process must be considered, such as products of incomplete combustion from agent destruction and toxic materials used in the disposal process. However, because the quantity of these toxic products is substantially smaller than the original quantity of agent, they generally represent a smaller hazard. Risks from toxic products are primarily initiated by internal events, such as process upsets. External risks are virtually nonexistent because they would usually result in a shutdown of the process. However, extreme external events could cause the release of hazardous materials, such as propane or sodium hydroxide, from on-site storage tanks.

Risk Receptors

There are three potential risk receptors: workers, the public, and the environment. Because of their proximity to the stockpile and agent processing operations, *workers* are at risk from the acute lethal (and nonlethal) hazards associated with agent releases, regardless of the initiating event (an at-risk situation). They are also potentially at risk from long-term exposure to very low concentrations (i.e., below the eight-hour time weighted average) of agent and the products and by-products of agent destruction. Workers are also susceptible to injury from ordinary industrial accidents (e.g., falls, burns, eye injuries, overheating in protective clothing), but these risks are not included in the risk assessments performed for the TOCDF. They can be better understood through safety inspections and analyses of injury rates and can be managed by following proven safety practices.

Risks to the *public* stem primarily from agent releases caused by external (catastrophic) events. The public could also be at risk from long-term exposure to the products and by-products of agent destruction, if they were released into the environment as a result of destruction processes.

Environmental risk is associated almost exclusively with the release of agent and the products and by-products of agent destruction to the environment.

Risk Measures

For humans (both workers and the public) there are three potential measures of risk either from the stockpile or from stockpile destruction: acute lethality; acute and latent noncancerous health effects; and latent cancer. The potential adverse consequences for the environment are the contamination of land and/or water and adverse effects on native or endangered species.

Risk Mitigation

The most effective mitigation of risk takes place before a hazardous material is released and is often called prevention rather than mitigation. However, after a hazardous material has been released, but before it reaches a receptor, risk mitigation is also possible, i.e., the consequences of the release can be reduced. Risk mitigation can include taking measures at the spill site (e.g., containing the spill), measures at the receptor site (e.g., using protective masks), and emergency response measures (e.g., shelters, evacuation, etc.). The QRA takes into account some of these measures. However, the primary intent of the QRA is to calculate a realistic estimate of risk to the public. The analysis is not structured to measure the effectiveness of the local Chemical Stockpile Emergency Preparedness Program (CSEPP). The QRA uses simple models and average data for mitigation (e.g., radial evacuation, evacuation time estimates for broadly defined conditions [time of day and weather, for example], using a representative evacuation speed of 8 m/sec (18 mph) and an assumption that 95 percent of the populace will take protective action). Representatives of the QRA team, the Army, and the CSEPP concluded that the QRA is the best estimate of risk. However, it may be more pessimistic than CSEPP calculations for some scenarios because the CSEPP uses more sophisticated models.

OBJECTIVES AND SCOPE OF THE DCD/TOCDF RISK ASSESSMENTS

Two separate risk assessments were performed for DCD/TOCDF. The first, a QRA, evaluated internal and external event-initiated risks to workers and the public. The second, an HRA, evaluated human health (public) and environmental risks associated with normal operation of the destruction facility. The committee's *Systemization* report (NRC, 1996b) observed that "the multiplicity of assessments can cause misunderstanding among reviewers, government agencies, and the public. The Army should adopt a standard language that recognizes the ensemble of risk-related projects as 'the risk assessment,' and individual studies should always be referred to as components of the wider 'risk assessment.'" The report advocated the preparation of a single TOCDF risk assessment summary report to present integrated QRA/HRA results.

The current report does not dwell on the risk assessment methodologies but deals explicitly with the DCD/TOCDF risk assessment and risk management processes. A more detailed description of the QRA methodology can be found in the *Systemization* report (NRC, 1996b). A very brief overview can be found in the following sections of this report. The HRA was performed following methodology described in public documents (EPA, 1994).

Quantitative Risk Assessment

The DCD/TOCDF QRA (U.S. Army, 1996c) had several objectives:

- to evaluate quantitatively the health risks from *accidental* releases of chemical agents to the public and to workers at the site
- to rank the plant and operational features at the facility that govern risk (Insights are to be used as a basis for risk management at the facility.)
- to compare the risks associated with the disposal process with the risks of continued storage (Insights are useful to the Army in making decisions regarding stockpile disposal, especially the specific order of items scheduled for disposal.)
- to provide a "living model" QRA that can be updated as changes are made to the facility or as additional insights into accident behavior become available (The living model will be one of the analytical tools supporting decision making within the TOCDF risk management program throughout the life cycle of the plant.)

The TOCDF QRA estimates the risk to the public and to workers from accidental releases of chemical agent associated with all activities during storage at DCD and throughout the disposal process at the TOCDF. Activities associated with the disposal process include:

- munitions storage at DCD prior to disposal
- munitions handling at DCD in preparation for transport to the disposal facility
- transport of munitions to the disposal facility
- the disposal processes

The study includes all identified potential causes of release, except for intentional acts, such as sabotage. Releases resulting from both internal initiating events (events that originate inside the facility or that directly result from activities during the disposal process) and external events (such as earthquakes, aircraft crashes, and tornadoes) are included.

Results of the TOCDF QRA are presented in terms of both public and worker risks. For public risks, both the risk of acute fatalities and the risk of exposure-induced cancer from accidents (mustard agents are potential carcinogens) are estimated. The risk of fatalities is presented in three ways:

- a risk profile showing the probability of exceeding a given number of deaths during the disposal period
- risk profiles as a function of distance from the site
- an average measure, e.g., the expected number of deaths during the disposal period

Appendix A of this report develops the bases for the presentations of risk (risk profiles and expected fatalities), explains how to interpret results, and discusses various measures for comparing risks. In estimating worker risk, the TOCDF QRA addresses only acute fatalities from accidents involving agent release caused by processing. Latent risks to workers from exposure were calculated but are not included in the results for several reasons:

- Workers directly involved in an accident are assumed to be killed, either from agent or from an explosion.
- Reporting the health effects for workers who are not directly involved, but who work in adjacent areas, would be deceptive for several reasons:
  - The model may not properly capture the close-in dose.
  - A convincing argument can also be made that projected latent effects from everyday activities (e.g., maintenance) are much greater than the latent effects from an agent accident. No

study of routine exposures has been done because no problem is apparent.
- The calculated latent risk to workers is very, very small compared to the acute risk.

The committee agrees that latent risks to workers from exposure to accidents is very small and including them in the QRA is not warranted.

Worker risk from continued storage was not assessed in the QRA because processing has already begun and activities related to disposal are of most interest to the Army. Worker risk from continued storage would require assessing limited worker populations and restricted activity schedules that no longer exist at DCD. Worker risks associated with industrial-type accidents also were not included in the QRA.

Uncertainty analyses showing the possible range of results, which were presented only for the public risks, incorporate the types of uncertainty discussed in Appendix A. All other risks were expressed as expected risk levels. The upper uncertainty bound shown for the QRA estimates is a measure of the analysts' confidence in the results. There is a 95 percent chance that the risk is less than the upper bound.

Health Risk Assessment

To complement the QRA and to meet Resource Conservation and Recovery Act (RCRA) permitting requirements, a screening-level HRA to estimate possible human health risks associated with exposure to airborne emissions from the TOCDF has been completed by the Utah Division of Solid and Hazardous Waste (DSHW) (Utah DSHW, 1996). The HRA also evaluated risks to wildlife and the environment. The scope of the HRA was limited to anticipated normal operating conditions with a fairly large allowance for emissions associated with operational process upsets. The HRA was a screening estimate in the sense that the results represent extreme upper bounds for normal and upset releases, well beyond the 95 percent upper bound described in the QRA.

APPROACH AND METHODOLOGY

Quantitative Risk Assessment

The Army completed and published the final report of the TOCDF QRA in 1996 (U.S. Army, 1996c). The TOCDF QRA, which was conducted following guidelines recommended by the NRC (NRC 1993b; NRC 1994b), quantitatively analyzes the probability and consequences of accidental releases of agent at the TOCDF facilities and the DCD storage area.

QRA Team

The DCD/TOCDF QRA was performed for the Army by SAIC. The major part of the QRA was performed by the SAIC analysts themselves (SAIC has strong in-house technical capability and extensive experience conducting QRAs for large-scale engineered systems). In areas where special expertise was required, external subcontractors or independent consultants were used. These areas included seismic hazards, structural mechanics, munitions fragility, and the latent health effects of agent. Operation of the QRA team was independent of the TOCDF site staff. However, the QRA team frequently interacted with the TOCDF staff to ensure the validity and completeness of the analysis.

Approach

The DCD/TOCDF QRA used the state-of-the-art approach to probabilistic-based risk assessment methodology that was first introduced to the nuclear industry in the 1970s in the WASH-1400 report (U.S. NRC, 1975). Since then, QRA methodology has gradually evolved into a sophisticated decision-support tool and is now well accepted and widely used to analyze complex engineering systems in the nuclear and chemical process industries (U.S. NRC, 1990; CCPS, 1989). Other approaches to risk assessment have been evaluated by the NRC and are more commonly used to assess health risks where assessing exposure depends heavily on dose-response characteristics (NRC, 1983, 1994c). Because of the complexity of DCD and TOCDF operations, and because conservative assumptions were made about agent lethality, the committee recommended the U.S. Nuclear Regulatory Commission method be used for the QRA (NRC, 1993b).

The QRA is based on a comprehensive set of logic models developed from the engineering design and operation of the disposal system and from various scenarios of potential system accidents. The risk at the site is then represented by the likelihood of these accident

SITE-SPECIFIC RISK ASSESSMENTS

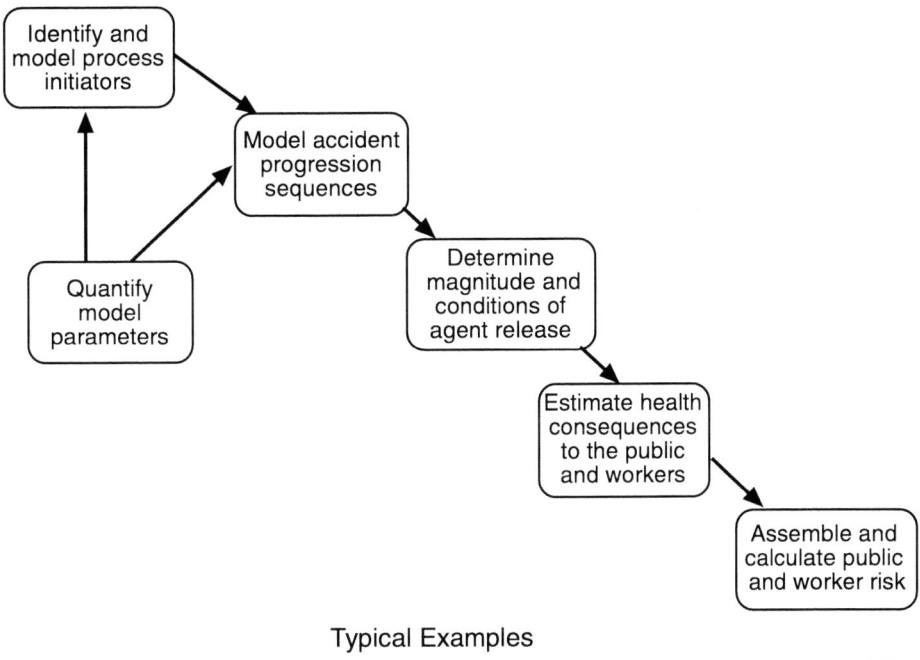

FIGURE 2-2 Overview of QRA process. Source: Adapted from U.S. Army, 1996d. Note: CHB/UPA means container handling building/unpack area.

scenarios and the severity of their consequences. The DCD/TOCDF QRA process is shown in Figure 2-2 and is summarized in the following paragraphs. Figure 2-2 also contains a table showing how a few elements of the QRA and HRA process are analyzed.

Identifying and Modeling Risk Initiators

The QRA starts with a systematic identification of deviations from normal process operations. These deviations are called "initiators." As suggested by the 1993 NRC letter report (NRC, 1993b), initiators considered in the DCD/TOCDF QRA include "internal" initiators, such as equipment failures and human errors, and "external" initiators, such as earthquakes, plant fires, floods, tornadoes, and aircraft crashes. Fault tree analyses, a well-accepted QRA logic modeling technique (Roberts et al., 1981), were used to identify the causes of internal initiators, a combination of equipment failures and human errors, for example.

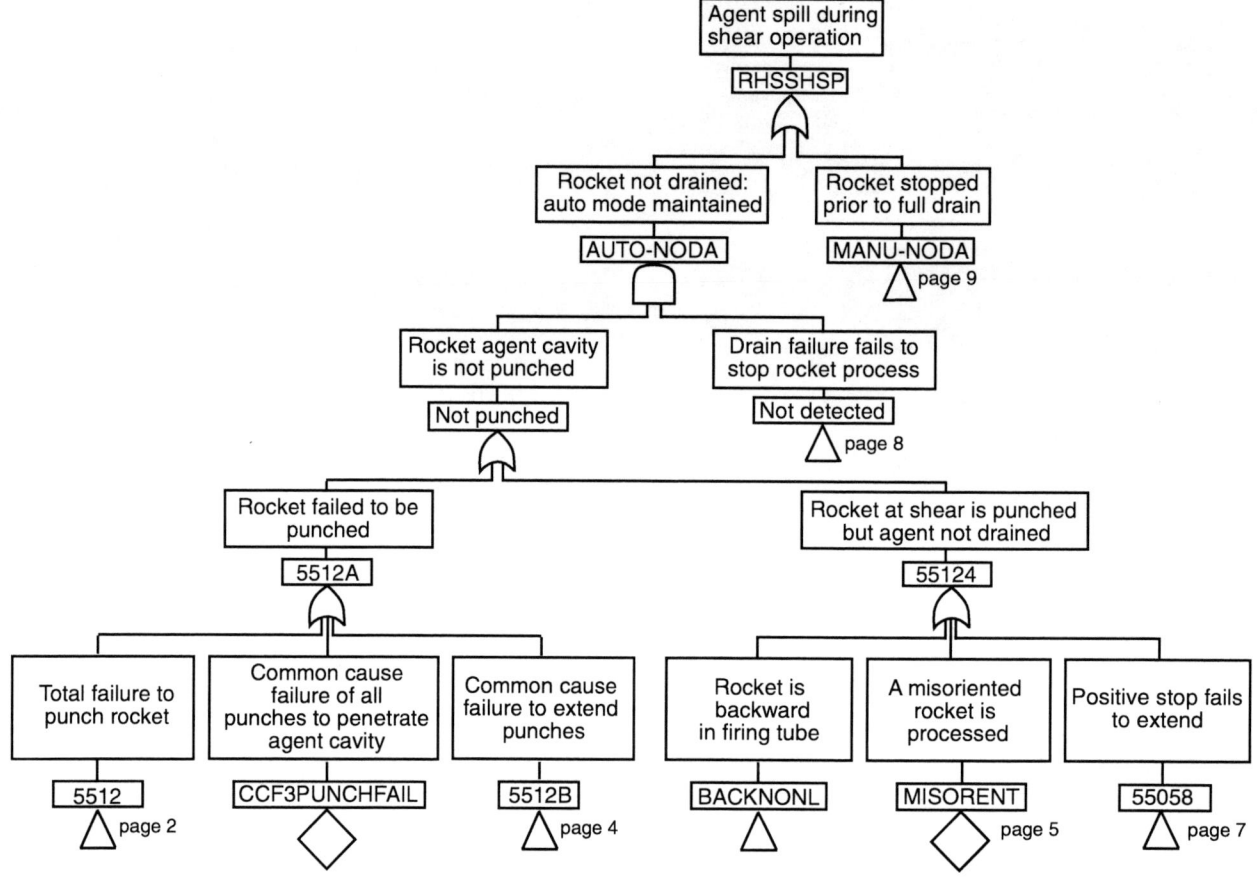

FIGURE 2-3 Rocket handling system fault trees for agent spilled during shear operation. Source: U.S. Army, 1996c.

To illustrate the method, the top logic from the DCD/TOCDF QRA fault tree for "Agent Spill during Shear Operation" is displayed in Figure 2-3, which has been extracted from an appendix of the QRA (U.S. Army, 1996c). This is one of many fault trees used to analyze operations at the facility. The rocket shear operation takes place inside a contained operating area (an enclosed room with a sealed environment to contain agent spills) and presents a hazard to workers in protective suits who would have to clean up any spills. Here the authors show that an agent spill during shearing operation can occur in one of two ways:

- The rocket is not drained (but automatic processing continues),
  OR
- The rocket is stopped before it is completely drained (and later is sent to the shear process).

The symbol with the pointed top and labeled RHSSHSP is called an "OR gate," which means that the event above it happens if either or both of the two events below it occur. The figure shows that the logic for the ways to "Rocket stopped prior to full drain" appears later in the fault tree, where one would learn that the draining is stopped if both of the following events occur:

- Processing stops before the rocket is fully drained, AND
- A human error is made (an operator sends an undrained rocket to shear).

An example of an "AND gate" in the figure can be found below the event "Rocket not drained; auto mode maintained." The AND gate has a rounded top and indicates that the event above it occurs only if all the events below it occur. In this case, "Rocket not drained, auto mode maintained" occurs if both "Rocket agent cavity is not punched" AND "Drain failure fails to stop rocket process" occur. The rest of the AND/OR logic for failure is easy to follow. On the bottom line of this figure is one new symbol, the diamonds below two of the event blocks, which indicate that the event will not be further decomposed. For example, "A misoriented rocket is processed" is an event that will be quantified directly, using available data.

Modeling Accident Progression Sequences

To specify potential accident sequences following an initiator, the TOCDF QRA uses the accident progression event tree (APET). Based on the engineering and operational information collected from the facility, the analysts identify and model (using the event trees to track different failure pathways) sequences of events following an initiator that could lead to an agent release.

Process accident flowchart models, called process operational diagrams (PODs) in the QRA, have been devised to encode process information. Upsets are identified in the POD. A POD is described in the *Tooele Chemical Agent Disposal Facility Quantitative Risk Assessment Methodology Manual* (U.S. Army, 1994):

> A POD is a step-by-step search for events and upsets...By asking a set of what-if questions after each successive operational step, a thorough assessment of potential upsets can be generated. During this process, existing analyses are referenced to ensure that previously suggested events are covered...[Start by] listing the major steps of the normal operations...Given each normal step, it is necessary to consider all deviations that could occur during that step or if that step did not happen properly...The PODs are used to document the steps in the process and allow efficient review by operational staff.

Quantifying Model Parameters

Data on equipment failures and human errors are collected from both industrial and CSDP experience and used as a basis for evaluating the likelihood of initiators as well as the likelihood of subsequent events leading to accidents and potential agent releases. The probability of accident sequences resulting in agent releases are then estimated based on the accident sequence model and the basic event data.

Determining the Magnitude and Conditions of an Agent Release

Following the identification in the APET of accident sequences that could lead to a release, the size of the release is estimated based on the event sequences. The amount of agent released and the conditions associated with the release are modeled for each accident sequence.

Estimating Health Consequences to the Public and Workers

Health effects to the public and workers are identified as the consequences of the accidental releases and have to be estimated. Mathematical models are used to estimate the dispersion of agent releases for site-specific weather conditions and to evaluate the exposure and resultant consequences to the public and to workers at the site. The Army's air dispersion code, D2PC (Whitacre et al., 1987; IEM, 1993), includes extensive chemical agent-specific data and models. However, it does not include statistical weather sampling, health effects models for agent exposure levels, population distribution modeling, or evacuation and sheltering models. The CHEMMACCS code (Haskin et al., 1995), developed at Sandia National Laboratory for use in QRA consequence analyses, uses the same agent-specific data and dispersion model as D2PC. The underlying Gaussian plume dispersion model is similar to the codes used by the Environmental Protection Agency (EPA). Input includes site-specific data for the TOCDF and the surrounding area.

Assembling and Calculating Public and Worker Risk

The risk of each accidental release is represented by the probability of the accident event sequence and the consequence of the release. The total risk from the disposal facility and the storage area is represented by combining risks from individual releases.

Presentation of Results

The DCD/TOCDF QRA provides detailed analyses of risk levels from several perspectives.

Risk Profile. A risk profile is a plot of the likelihood of "x or more fatalities" plotted as a function of "x." The uncertainty bounds (at least the 5th and 95th percentile) are usually included for calculated mean and median values. The *mean* value represents an average of the range of estimates. The *median* value is the point at which half the estimates are higher and half lower.

Expected Fatalities or Expected Number of Induced Cancers. The expected number of fatalities or the expected number of induced cancers are statistical summations over all impact scenarios of the individual expected values (the product of the probability of an individual accident release and its consequence in terms of expected fatalities). This number is most often referred to as the *risk value*.

Dominant Contributors to risk. Dominant contributors are sequences of events that rank high in risk value. Effective site-specific risk management would identify dominant contributors in seeking the most effective ways to reduce risk within the CSDP.

Health Risk Assessment

The HRA performed by the Utah DSHW followed guidelines and methods that have been established and prescribed by the EPA and used for assessing the acceptability of a broad range of health and environmental risks (EPA, 1994).

Screening-level risk assessments are conservative by design in that they are based upon worst-case assumptions when operational data are not available. Because TOCDF was a new facility with no operational history at the time the HRA was prepared, a great many default assumptions were used. Because site-specific values were not available for wind speed profiling exponents, terrain adjustment factors, surface roughness, and scavenging coefficients, default values were used. Site-specific inputs were confined to local geographical, hydrological, meteorological, and agricultural information. Site emissions data were approximated based on data from JACADS operational experience because the TOCDF had not yet begun operations.

For the HRA, six point sources of emissions are identified, including the TOCDF incinerators and two other areas where the products of the agent destruction process are handled. A description of the facilities is given in the *Recommendations* report (NRC, 1994b). The six point sources included in the HRA are:

- liquid incinerators
- metal parts furnace
- deactivation furnace system
- dunnage incinerator
- brine reduction area stack
- heating, ventilation, and air conditioning filter stack

The potential impact from each point source was evaluated.

Impacts from the TOCDF combined stack (the metal parts furnace, liquid incinerators, and deactivation furnace system in simultaneous operation), from all sources at maximum TOCDF operations, as well as from combined TOCDF and CAMDS operations, were considered.

The HRA identified four categories of constituents of potential TOCDF emissions: chemical agents, metals, and volatile and nonvolatile agent decomposition products (i.e., products of incomplete combustion). Sixty individual constituents were identified.

Human exposures were considered to occur both directly and indirectly. The inhalation of emissions (direct) as well as the ingestion of contaminated soil and food (indirect) were exposure mechanisms deemed appropriate for purposes of the HRA.

Consistent with EPA guidelines for screening-level risk assessments, an adult resident, a child resident, a subsistence fisher, and three different subsistence farmers were identified as likely receptors. The adult and child residents were considered to reside at the off-site point of maximum emissions impact. The subsistence fisher was located in an area where subsistence fishing was thought to be practiced, and the three subsistence farmers were located based upon a survey of farming in the area. The HRA considered potential human health risks based on scenarios of 10, 15, and 30 years of continuous TOCDF/CAMDS operation, although the TOCDF is scheduled to operate for only 7.1 years.

STOCKPILE COMMITTEE OVERSIGHT

As a standing committee of the NRC, the Stockpile Committee reviewed the technical developments that led to the design of JACADS and the first risk assessment in the FPEIS (U.S. Army, 1988). That review and concern about the need to understand the risk at each site led to the committee's letter report on risk assessment (NRC, 1993b), which essentially laid out a specification for site-specific risk assessments. Later the committee's *Recommendations* report (NRC, 1994b) further defined that specification and reiterated the need for site-specific risk assessments.

As the Army's site-specific risk assessment for DCD/TOCDF took shape, the committee's role was defined as oversight of the risk assessment/risk management process and oversight of the Expert Panel review process. Actual detailed technical review of the QRA was the charge of the independent panel of experts in risk assessment and chemical engineering (information on the members of the panel is given in Appendix B). However, the committee took advantage of many opportunities to examine the technical details of the risk assessment work. The committee's *Systemization* report (NRC, 1996b) reviewed the methodology of the QRA and the efforts of the Expert Panel. The committee found that:

- The QRA methods met the recommendations of the committee's earlier reports.
- The SAIC QRA team was being responsive to committee questions and Expert Panel comments; they were developing new analysis tools for first-of-a-kind QRA calculations, retaining outside expert groups in difficult technical areas to advise them in areas where the literature was incomplete, conducting tests and new mechanistic analyses to answer new technical questions, and revising the QRA analyses based on new information.

Thus the QRA was being modified and extended as work progressed to respond to advice from the Expert Panel and prior recommendations of the committee. Furthermore, answers to questions sometimes required revised approaches to specific aspects of the QRA analysis.

Quantitative Risk Assessment

The committee closely followed the risk assessment activities in three ways. The Army and its contractors made presentations on the technical progress of the QRA at all regular quarterly meetings of the committee and at some special meetings. During these sessions, the QRA team responded to detailed questions from the committee. In addition, two members of the committee attended the meetings of the Expert Panel, observing the process and also having the opportunity to question the analysts actually involved in all aspects of the QRA. Finally, all members of the committee received the draft of the Main Report of the QRA to review, and three members received the entire report including the extensive appendices. Questions generated in this process were raised at subsequent Expert Panel meetings.

Health Risk Assessment

The committee received the protocols under which the HRA was to be performed. It also received a briefing by the state of Utah and its consultants on the HRA results, assumptions, and models during a regular quarterly committee meeting. Three members of the committee received the HRA document. The HRA was performed in accordance with EPA and Army protocols because the HRA is required to meet legal requirements and must be done in accordance with standard methods. The Army and the state of Utah agreed to perform an EPA-style HRA using conservative worst-case analysis rather than best-estimate and uncertainty analysis. The HRA showed that the risk is low and meets the permitting requirements so no special risk management efforts are required for normal and mild upset conditions. It is also clear that the risk of accidents is much higher than the risks examined in the HRA. Therefore, the committee can find no compelling reason for the Army to extend the HRA for the purpose of directly combining and comparing the results of the two studies.

ADDITIONAL REVIEW OF THE RISK ASSESSMENTS

In addition to the Stockpile Committee, other organizations have been involved in the review and guidance of the risk assessments.

Quantitative Risk Assessment

Three principal reviews were used throughout the development of the QRA: intraproject reviews, PMCD and TOCDF reviews, and independent external reviews.

Intraproject Reviews

Intraproject reviews were conducted according to the quality assurance requirements established by the PMCD. Analysis and integration models and results at the subtask level, the task level, the integration level, and the assembly level were reviewed by SAIC analysts and engineers with experience performing QRAs on large-scale integrated engineering systems.

PMCD and TOCDF Reviews

The PMCD and TOCDF management provided most of the engineering and operational data used in the DCD/TOCDF QRA. They also reviewed the QRA models and results to ensure that the facility was correctly modeled in the QRA. The operational diagrams and models developed by the QRA team to analyze potential accidents at the TOCDF were reviewed by field engineers familiar with the TOCDF processes. As failure sequences were modeled during TOCDF systemization, PMCD and site personnel were asked if the results were consistent with their general knowledge and operating experience at JACADS. They were also asked to brainstorm on types of failures that might have been omitted.

Integration of the PMCD and TOCDF reviews into the QRA process at an early stage led to the establishment of an effective communication network. This not only kept the QRA team analysts well informed of the status of the facilities and of ongoing activities, but also helped the PMCD and TOCDF staff understand the risks associated with the disposal processes and the significance of the risk assessments being done as early as possible in the project.

Independent External Reviews

The PMCD also established the Expert Panel (see Appendix B) through a separate contractor, MITRETEK Systems, to oversee the conduct of the QRA. This independent review group consisted of five experts, each of whom was either a specialist in the QRA field, a professional in the chemical industry, or an expert who specialized in chemical process safety. The Expert Panel, which met for the first time in November 1994 and regularly thereafter, followed the progress of the QRA through regular, interactive meetings with the project team. The Expert Panel had full access to all analytical activities and maintained an ongoing dialogue with the QRA team. Representatives of the Stockpile Committee attended the second panel meeting in February 1995 (as observers) and have attended all panel meetings since then.

The Expert Panel reviewed and evaluated the QRA methodology, data, procedures, and assumptions. On March 28, 1996, the panel briefed the Stockpile Committee and indicated that the panel members were, in general, satisfied with the QRA methodology. The panel indicated that the SAIC study extended the state of the art in several areas (MITRETEK Systems, 1996). The panel also pointed out that the QRA analysts responded positively to comments. Appendix S of the QRA provides documentation of comments from the Expert Panel and responses of the QRA analysts (U.S. Army, 1996c). The final report of the Expert Panel is now available (MITRETEK Systems, 1996).

The committee concurs with the Expert Panel's findings (MITRETEK Systems, 1996):

- The methodology was sound and has extended the state of the art in several areas.
- The methodology was well implemented.
- The panel had some reservations about a few technical aspects of the QRA but was reasonably satisfied that these did not affect the overall conclusions.

The committee notes that the panel had a significant impact on several key areas of the QRA:

- The treatment of uncertainty is now more clearly addressed.
- The seismic vulnerability analysis for the liquid propane gas tank has been improved.
- The model for workers donning masks after a strong earthquake is more realistic.
- The mechanistic modeling of munitions handling accidents is much improved.
- The interactive independent review process was effective.

Significant improvements in the QRA methodology have been made.

SITE-SPECIFIC RISK ASSESSMENTS

Health Risk Assessment

The HRA was prepared by the Utah DSHW following procedures established by the EPA. The TOCDF HRA was issued by the DSHW for public comment before publication of the final version and before acceptance of the HRA as part of the TOCDF RCRA operating permit.

RESULTS

Quantitative Risk Assessment

In most QRA studies at least two classes of consequences are considered—acute and latent health effects. Acute health effects involve immediate injuries and deaths. The immediate injuries associated with agent release at the TOCDF tend to be minor reversible effects from very low level exposures to nerve agent (e.g., watery eyes and runny noses). In comparison to deaths and latent cancer effects, immediate injuries are minor and are not reported in the DCD/TOCDF QRA. The most severe latent health effects are possible cancers from exposure to mustard. These cancers, if not properly treated, can become deaths many years later. The risk profiles that follow are associated with immediate (acute) fatalities.

Stockpile Risk to the Public

The dominant risk-initiating event for storage of the chemical stockpile at DCD is an earthquake. Although earthquakes are infrequent, they have widespread effects and could cause the release of much more chemical agent than other types of accidents. Seismic events contributing to the risk have mean accelerations above 0.2 g and recurrence intervals of 1,000 years or more. Such earthquakes significantly exceed normal building code design values and thus can lead to failures of equipment and structures. Overall, earthquake initiated events account for 82 percent of the average public fatality risk associated with continued storage of the stockpile; of the 18 percent nonseismic public fatality risk, leaks of agent GB from ton containers account for 11 percent (Figure 2-4).

An aircraft crash into storage structures and the electromagnetic effects of lightning (which could cause a

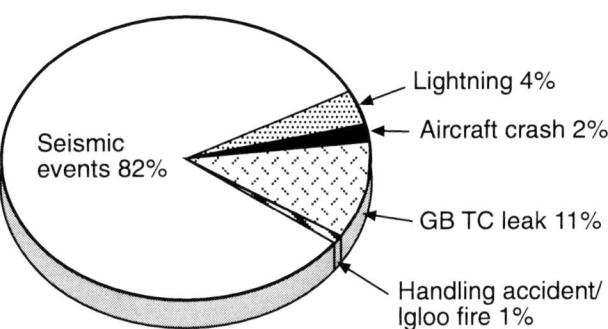

FIGURE 2-4 Contributors to the average public fatality risk from continued storage at DCD. Source: Adapted from U.S. Army, 1996d.

fire in an igloo or cause an M55 rocket to ignite) were also considered. The results shown in Figure 2-4 indicate that the impact of these initiators is only 2 percent and 4 percent, respectively, of the total storage risk. Risks from reconfiguring associated with normal storage maintenance, such as isolating leaking munitions, account for about 1 percent of the storage risk. These maintenance activities are infrequent, and the potential for a significant release is small because the number of munitions handled at any given time is limited. For this reason, and because destruction operations are already under way, the risk to workers from continued storage was not included in the QRA.

Figure 2-5 shows the public acute risk profile associated with stockpile storage at DCD during the disposal processing period. (Refer to Appendix A for an explanation of risk profile curves.) Uncertainty bounds are indicated. The uncertainty calculations for the QRA are given in terms of uncertainty in the probability of exceedance. For example, if the question is the probability (chance) of 100 or more fatalities, a vertical line can be constructed at 100 fatalities showing that the median (50/50) probability is 5×10^{-8} or 1 in 20 million, the lower bound shown (5th percentile) is 5×10^{-9} or 1 in 200 million, and the upper bound shown (95th percentile) is 3×10^{-6} or 1 in 333,000. The mean (average) value over this distribution is 6×10^{-7} or 1 in 1.7 million. Uncertainties exist for all of the risks discussed in this report; Figure 2-5 shows the general magnitudes of the uncertainties. Many of the QRA results elsewhere in the report are shown as mean risk values to simplify presentation, but all calculations include uncertainties.

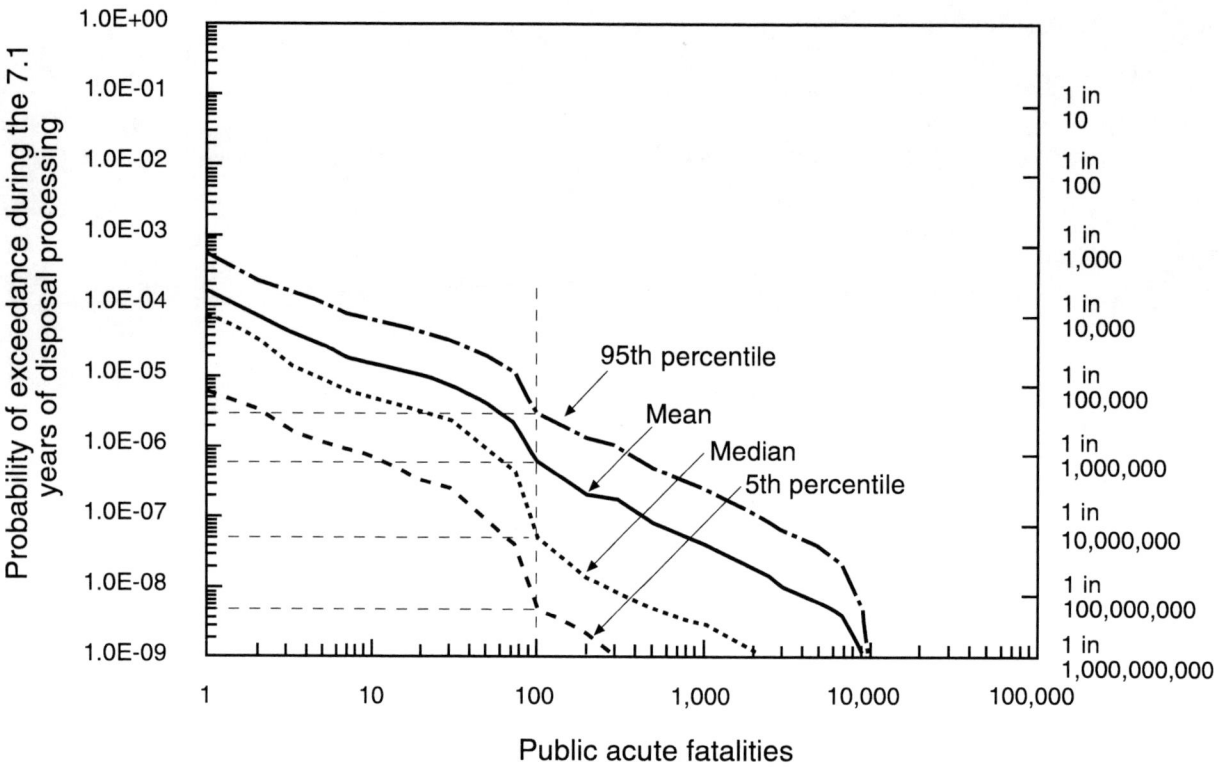

FIGURE 2-5 Public acute fatality risk of DCD stockpile storage over 7.1 years of disposal processing. Source: Adapted from U.S. Army, 1996c.

In a more detailed view of the results, it is possible to consider the risk as a function of distance from the site. Figure 2-6 shows distances surrounding the site in concentric rings. The QRA analysis is based on actual population densities around the site. Figure 2-7 shows the variation in risk for the concentric rings surrounding the disposal facility. The contribution from each ring is a component of the overall mean risk profile shown in Figure 2-5. Note that the close-in rings dominate the risk at the lower number of fatalities. For example, the mean probability of one or more fatalities (1.6×10^{-4} from Figure 2-5, a result of summing the probabilities for all the rings) comes primarily from the 2 to 5 km ring (1.0×10^{-4} or 1 chance in 10,000). This is because the chance of exposure is much higher closer to the site. To reach large numbers of people, the plume must travel to larger population centers. The chance of 10 or more fatalities is 1.5×10^{-5} (less than 1 in 100,000 from each of the two nearest rings, 2 to 5 km and 5 to 10 km). Risks of one or more fatalities during disposal operations fall below 1 in 1 million beyond 15 km from the site; for stockpile storage risks, the 1 in 1 million risk zone extends beyond 35 km.

As the stockpile continues to age and agent containment systems deteriorate, risks can be expected to increase. However, current calculations, based on observed degradation, indicate that no significant increase in annual public risk is expected in the next 20 years. Therefore, the storage analysis in the QRA assumes that the stockpile will not degrade in the next 20 years.

Operational Risk to the Public

The best way to evaluate the public risk of processing is to compare that risk with the risk of continued storage. This comparison is made in Figures 2-8 and 2-9. All risks on these curves are shown on a per-year basis so that they are directly comparable. Consider Figure 2-8, where the risks of continued storage, assuming no processing takes place, are indicated by the broken line. The vertical axis shows average fatality risk per year, and the horizontal axis shows the time line for disposal. The risk level for the first GB disposal is only about 0.00006 fatalities per year with a processing duration of about nine months. Note, however, that stockpile storage risk decreases at the end of

SITE-SPECIFIC RISK ASSESSMENTS

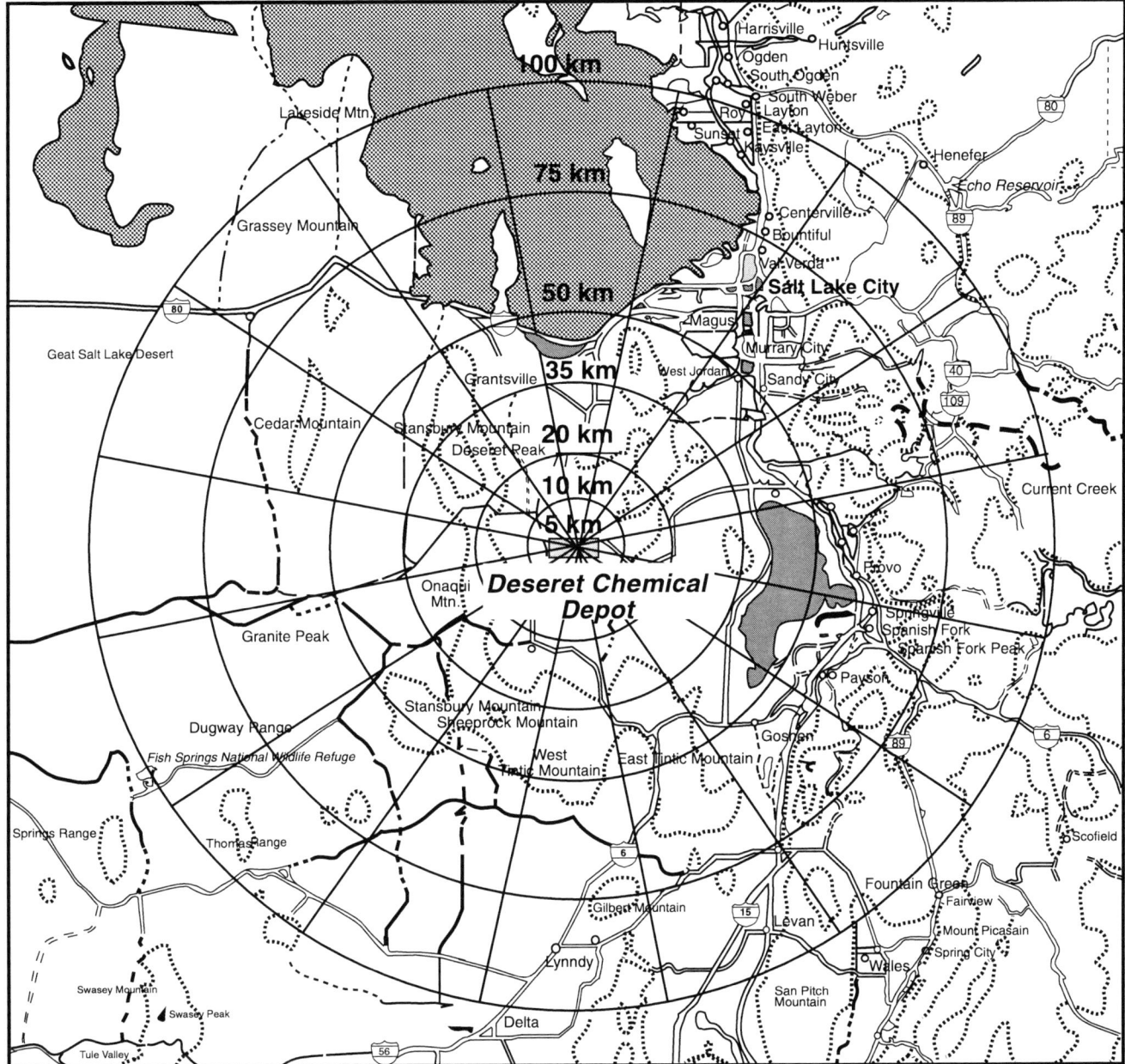

FIGURE 2-6 Radial polar grid of surrounding population. Source: U.S. Army, 1996c.

that time by two-thirds because the most risky items would be removed from the stockpile during the first disposal campaign.

By the end of the fifth campaign (GB ton containers nearly three years into disposal operations), both the storage and the processing risk have essentially disappeared. Nevertheless, although the risk is small, it is clear that storage risk is still much greater than processing risk and that accepting the processing risk for three years dramatically reduces the total risk.

Figure 2-9 presents the identical information on a logarithmic scale. Risk analysts and managers like this presentation because it emphasizes details that cannot be seen on the ordinary, or linear, scale of Figure 2-8. To others, Figure 2-9 may appear to distort the information and overemphasize very low levels of risk. Using Figure 2-9, risk managers at the TOCDF were able to ascertain the relative effects of various agent destruction campaigns. This information was used to reorder the disposal campaigns to minimize the total overall risk.

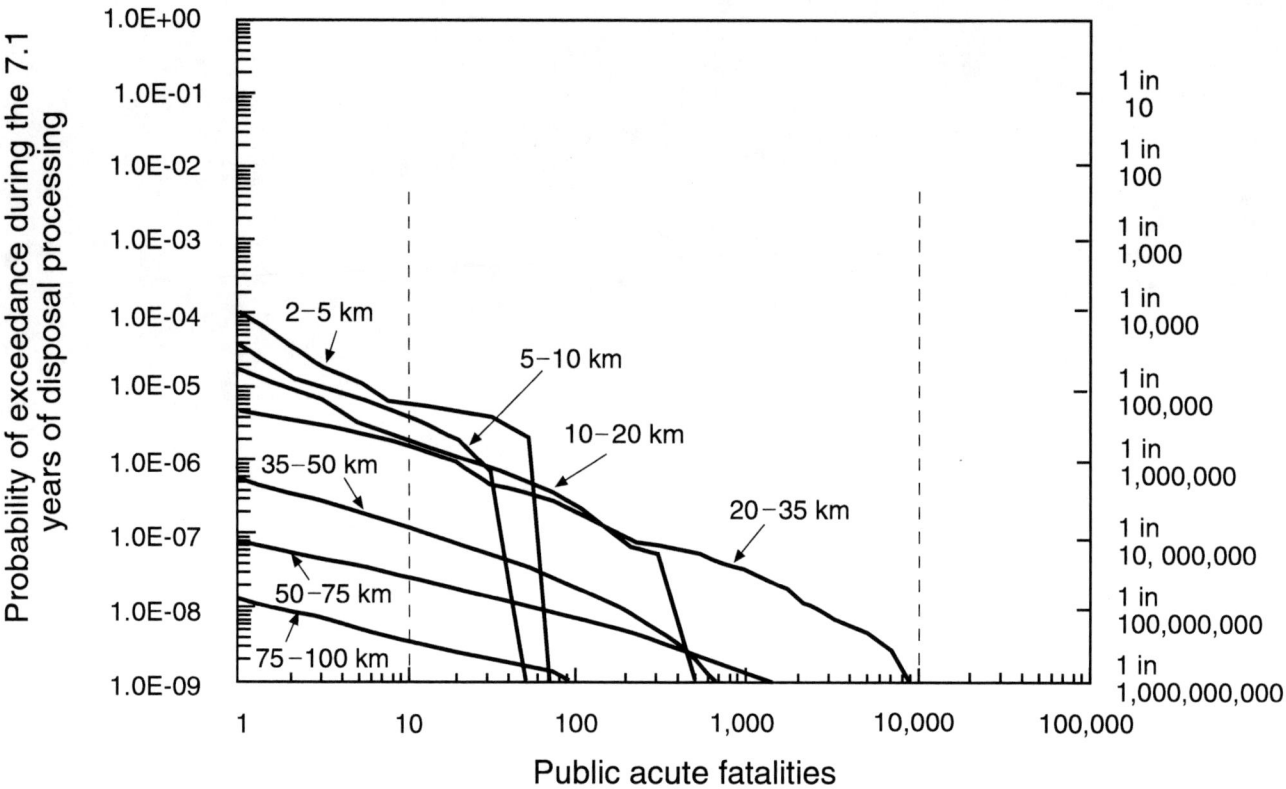

FIGURE 2-7 Mean public acute fatality complementary cumulative distribution function for munition storage during the 7.1 years of disposal processing, by distance from DCD. Source: Adapted from U.S. Army, 1996c.

For disposal processing at the TOCDF, the QRA results show that public fatality risk is dominated by earthquakes (97.4 percent) as the most dangerous risk-initiating event (Figure 2-10). The consequences of an earthquake at the TOCDF are further dominated by the potential for a structural failure in the unpack area of the container handling building area caused by an earthquake stronger than the building is designed to withstand. The severe consequences would result partly because munitions are unpacked in this area and are not protected by transport containers.

The QRA results shown in Figure 2-10 also indicate that internal events associated with processing account for less than 1 percent (i.e., 0.8 percent) of the TOCDF risk and that nearly all of this risk is associated with handling rather than with actual agent destruction. The study credits the low risk of processing to the safety and mitigation features of the baseline system and the limited quantities of agent available for release during processing.

Operational Risk to Workers

Workers at the TOCDF, including all support and administrative staff located at the facility or in nearby buildings and munition handlers responsible for removing munitions from the stockpile and transporting them to the disposal facility, were included in the risk assessment. The study includes only worker risks associated with accidents involving agent releases. Processing and handling workers can be directly affected by the blast of an explosion, for example, or by agent dispersion from an accident, and both effects are included. However, industrial-type risks, e.g., being crushed by a lift-truck, were not considered. The QRA results indicate a 1 in 7 probability of a worker fatality in the total disposal-related worker population in the 7.1 years of disposal processing. Figure 2-11 shows the contributions of various causes to worker risk. Maintenance activities account for 44 percent of the risk; seismic events, 36 percent; metal parts furnace explosions, 6 percent; handling, 6 percent; and other causes, 8 percent.

The QRA indicates that risks to disposal-workers from agent-related accidents are substantially higher than the public risks, as would be expected because of the proximity of the workers to the agent. Small releases that would not have an impact at a significant distance could still be lethal to workers in the immediate area. According to the QRA, there are about 500 workers at the TOCDF. If the 0.13 expected fatalities per 7.1 years of operation are dominated by single fatality accidents, then the individual disposal-worker risk at the TOCDF is about 4×10^{-5} per year. By comparison, this risk is about equal to the total occupational risk for all occupations (based on 1995 Occupational Safety and Health Administration [OSHA] data for fatalities by occupation) (U.S. Department of Labor, 1995). Higher risk levels are encountered by workers in the construction industry (about 12×10^{-5} per year). The individual risk levels for the general population of operators, fabricators, and laborers is about 11×10^{-5} per year. Presumably, most of the TOCDF workers would bear this sort of job-related risk plus the agent-related risk, which would increase the level of risk by about one-third.

The OSHA data are averages across a wide spectrum of work environments and do not represent the lower individual risk levels that can be achieved by companies that emphasize safety. Because the disposal-worker risk levels from agent exposure are added to the normal occupational risk level, the committee believes that emphasizing job safety, both for agent and non-agent activities, is very important.

The risk for other on-site workers (outside the TOCDF and DCD storage area) is evaluated in the same manner as public risk. The probability of one or more fatalities for other on-site workers during the 7.1 years of disposal processing is 5×10^{-4} (1 in 2,000). With about 100 workers in this category, and assuming that most accidents cause a single fatality, the

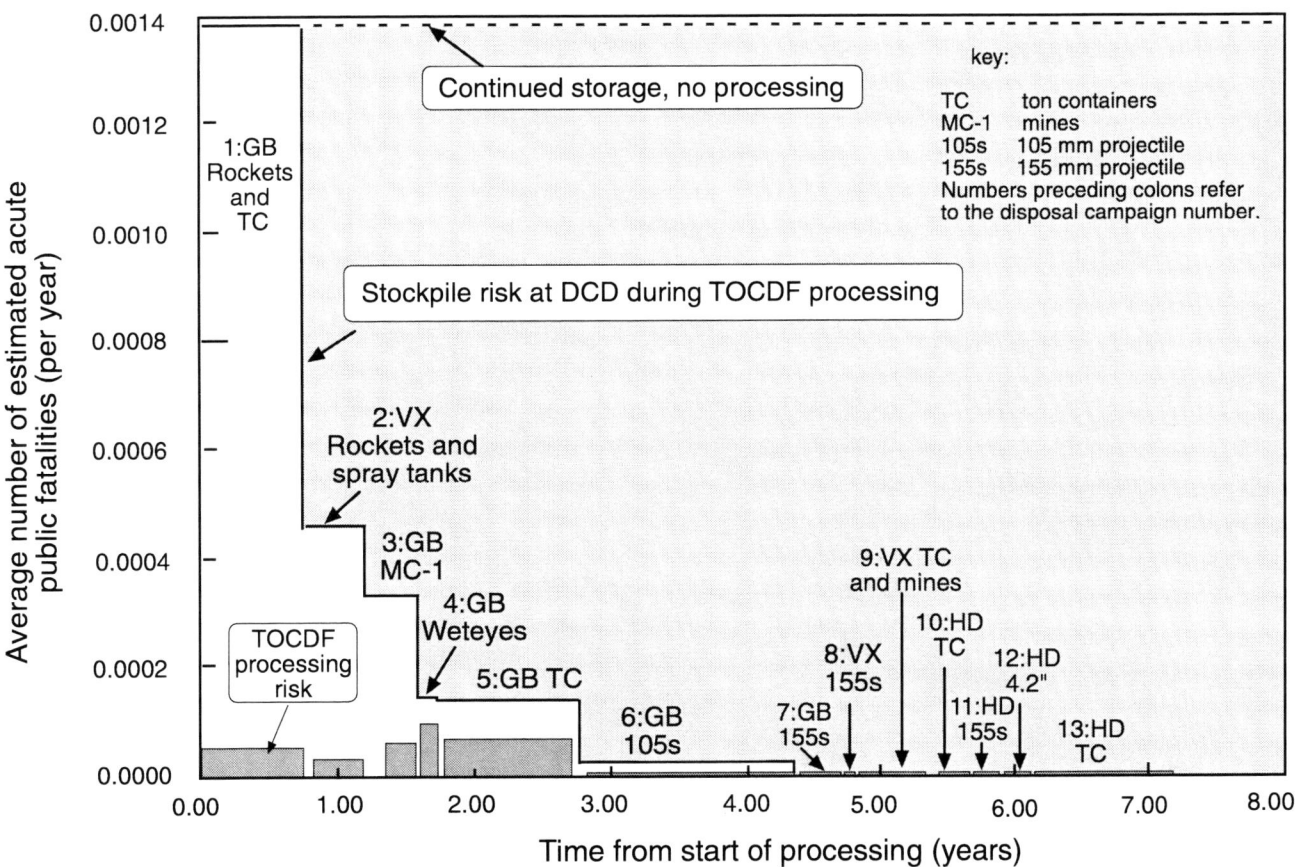

FIGURE 2-8 Comparison of public risks during processing at DCD and TOCDF. Source: Adapted from U.S. Army, 1996d.

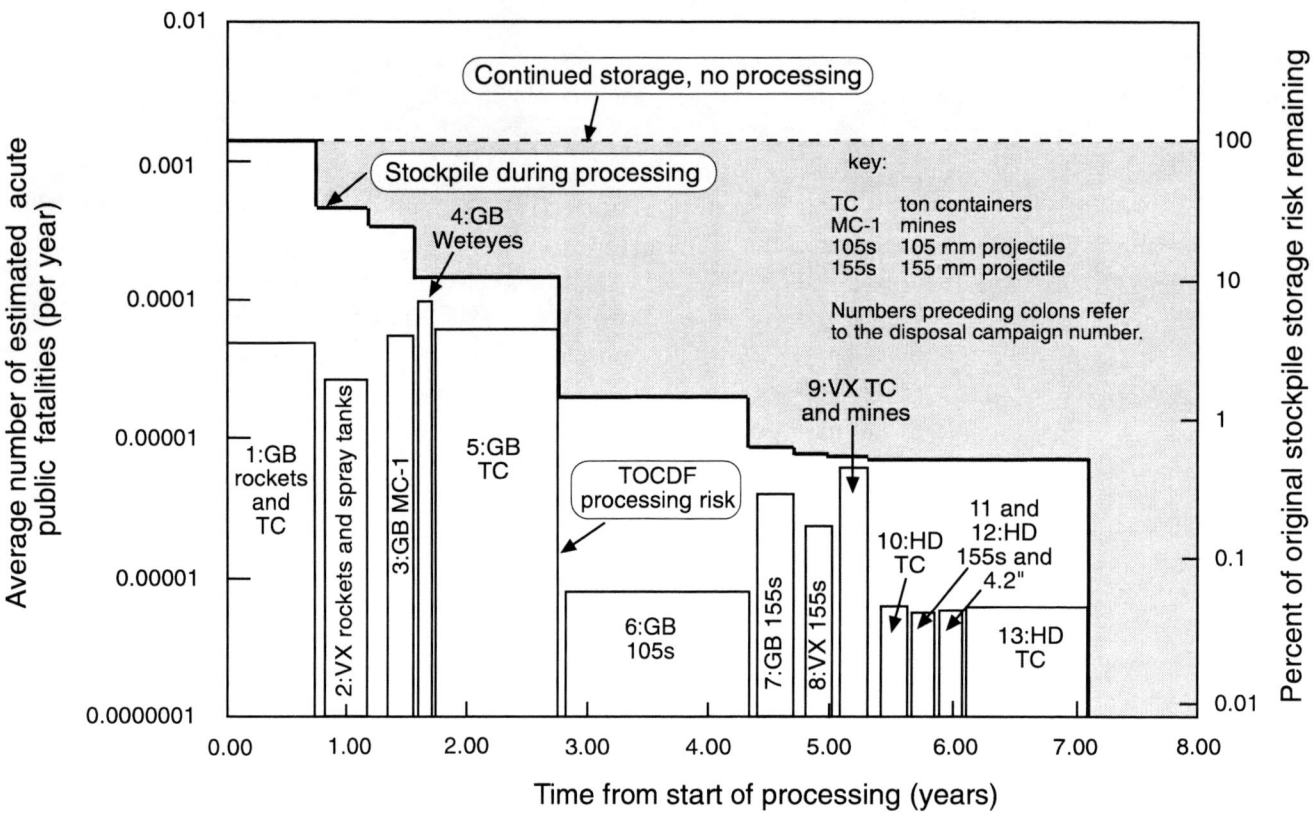

FIGURE 2-9 Comparison of public risks during processing at DCD and TOCDF (logarithmic scale). Source: Adapted from U.S. Army, 1996d.

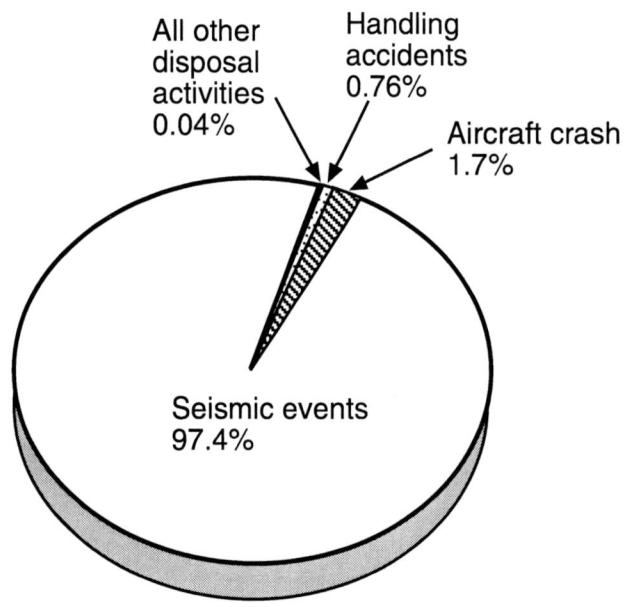

FIGURE 2-10 Contributors to the average public fatality risk from processing at DCD and TOCDF. Source: Adapted from U.S. Army, 1996d

individual annual risk is less than 1×10^{-6} (1 in 1 million per year) for other on-site workers. This risk is small in comparison to risk levels in standard occupations (on the order of 1×10^{-4} per year). Thus other on-site workers are not significantly affected by the movement and disposal operations at DCD/TOCDF.

Overall Risk

Acute Fatality Risk. The public risk of an acute fatal poisoning from agent release is shown for DCD and the TOCDF in Figure 2-12 as a risk profile. The vertical axis shows the probability of a release at the site in which the number of fatalities would equal or exceed the number on the horizontal axis. The risks associated with the three situations of concern are summarized in three curves: the risk of disposal processing at the TOCDF; the risk of storage at DCD during the disposal process (with allowance for depletion of the stockpile during disposal); and the risk of continued munitions

SITE-SPECIFIC RISK ASSESSMENTS

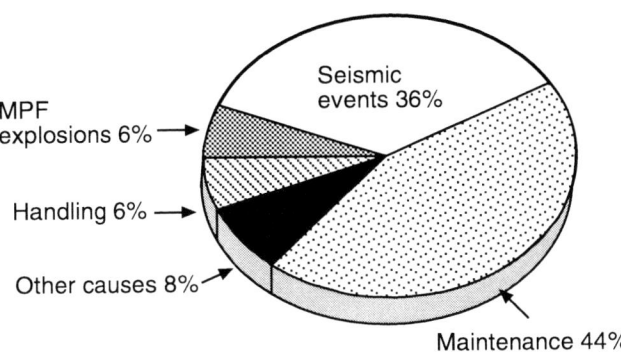

FIGURE 2-11 Contributors to the average risk of fatality to disposal-related workers at DCD and TOCDF. Source: Adapted from U.S. Army, 1996d.

storage (without disposal) at DCD. The analysis assumes that the disposal process will last approximately 7.1 years. The risk for 20 years of continued storage considers the case of a 20-year delay in initiating disposal operations.

According to Figure 2-12, the average probability of incurring one or more public fatalities is about 1×10^{-5} (approximately 1 in 100,000) for the 7.1 years of disposal processing at the TOCDF; about 1.4×10^{-4} or (1 in 7,000) for stockpile storage at DCD during the disposal period; and about 5×10^{-3} (1 in 200) for continued stockpile storage at DCD for the next 20 years, with no processing. Figure 2-13 shows the same mean risk profile for disposal processing for 7.1 years of TOCDF operations with uncertainty bounds.

The expected number of public fatalities over the time period is the probability-weighted sum of each possible number of fatalities (see Appendix A). The results of the QRA indicate that the expected number of fatalities is approximately 0.00016 for the 7.1 year disposal processing period, 0.002 for the stockpile storage at DCD during disposal processing, and 0.03 for continued stockpile storage at DCD for 20 years. The total expected public acute fatalities during the disposal operations is the summation of both the processing risk

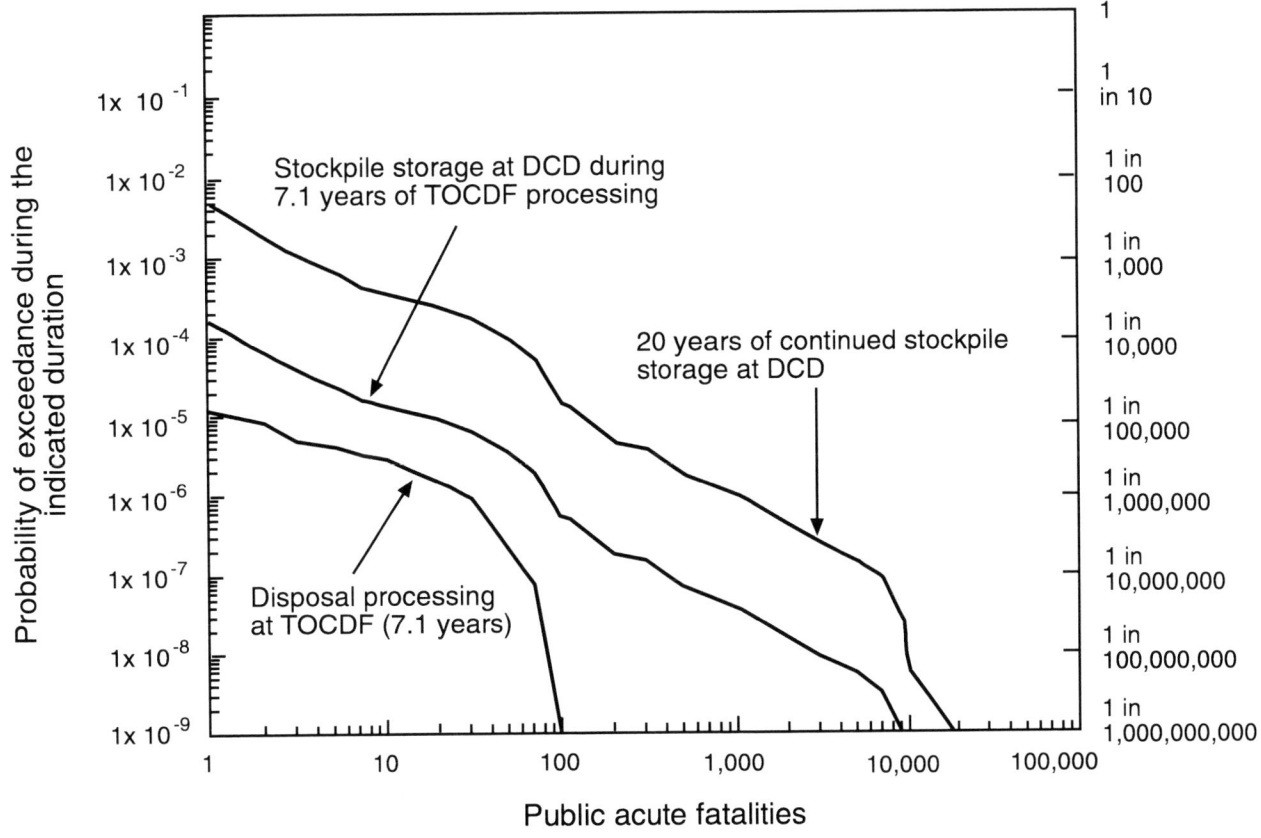

FIGURE 2-12 Summary of mean public risk from storage and processing at DCD and TOCDF. Source: Adapted from U.S. Army, 1996d.

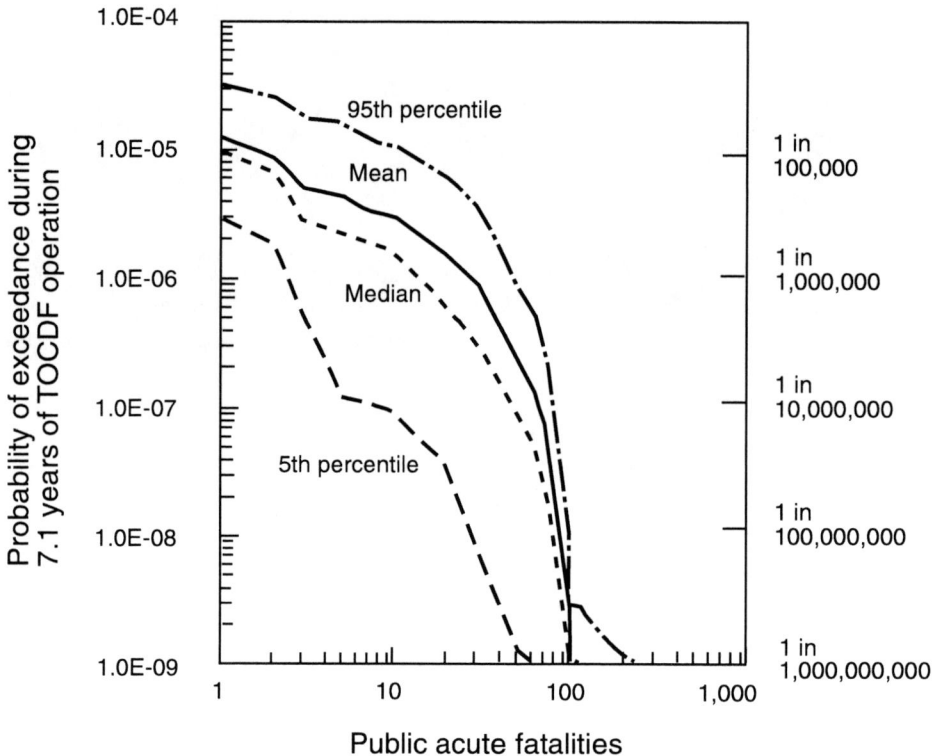

FIGURE 2-13 Public societal acute fatalities for all campaigns (TOCDF disposal processing). Source: Adapted from U.S. Army, 1996c.

and the storage risk during processing. This probability is only slightly higher (0.0021 versus 0.0020) than the storage risk alone because the processing risk is relatively small in comparison with the storage risk.

As anticipated, the risk decreases with distance from the site. Figure 2-14 shows how disposal processing risk profiles vary with distance from the site. The distances reported are measured from the main processing building. Since the TOCDF is not very large (approximately 250 yards × 300 yards), these distances are essentially the same as the distance from the TOCDF fence. Risk associated with the disposal process also varies with time, as the various agents and items are sequentially destroyed. Figures 2-8 and 2-9 show these variations as the disposal processing moves from one agent and munition configuration to another. The disposal sequence was adjusted after an early QRA draft showed it would be desirable to eliminate the most hazardous agents (GB and VX) earlier in the program. More than 90 percent of the accidental risk at the Tooele site (storage and processing) is associated with agent GB, largely because of its higher evaporation rate in comparison to the more acutely toxic, but low-volatility, agent VX. Accident sequences involving VX contribute about 10 percent of the overall public risk. The risk from a release of mustard is very small in comparison. An aircraft crash into the mustard ton container storage area and an ensuing fire pose the greatest threat from mustard.

The public risk is dominated by seismic initiating events. In the absence of an earthquake-initiated event, the mean fatality risk would be 40-fold less for the public, 10-fold less for other (nondisposal) on-site workers, and 16-fold less for disposal-related workers.

Worker Risk. Figure 2-15 illustrates the risk profile for other on-site workers at DCD/TOCDF. Risk profiles for disposal-related workers are not provided because current methods are not capable of combining the remote (dispersed agent-related) and direct (explosion, close-in agent, etc.) effects. The remote agent effects include variations in weather conditions and other external factors. The direct effects are calculated on a scenario-by-scenario basis. The QRA team could have produced risk profiles for remote effects alone but believes that they would have been incomplete. The committee agrees.

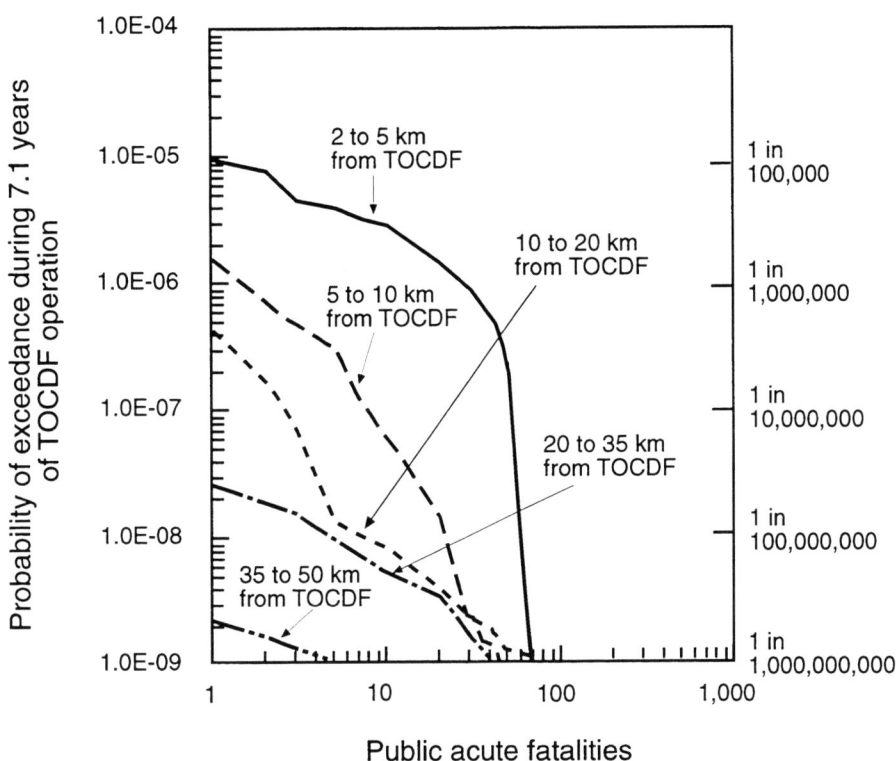

FIGURE 2-14 Mean public acute fatality risk by distance from TOCDF during disposal processing. Source: Adapted from U.S. Army, 1996c.

Public Cancer Risk. There is essentially no risk of cancer from accidental releases during processing or during 20 years of continued storage. The probability of an individual developing cancer is vanishingly small. The mean cancer risk to the public (the expected number of excess cancers) from storage for 7.1 years is only 0.000002; the mean cancer risk from processing is only 1 percent of the risk from storage. These are small risks in comparison with the acute fatality risk to the public associated with the facility.

Health Risk Assessment

To evaluate human health risks, both carcinogenic and noncarcinogenic health effects from chemical agents, metals, volatile and semivolatile products of incomplete combustion, and other combustion products were considered in the HRA screening risk assessment. Because carcinogenic effects dominate public concerns, the following section focuses on this aspect of the HRA. According to EPA guidelines (EPA, 1994), to protect human health and the environment, emissions of carcinogens should not exceed a cancer risk level of 1×10^{-5} to a maximally exposed individual over a 70-year lifetime. This corresponds to a 1 in 100,000 chance of developing cancer from exposure under a particular scenario. The selection of this level acknowledges that background exposures from drinking water, food, and air, that is, sources other than the emission source being evaluated in the screening assessment, also contribute to the risk in the study area. Setting the cancer risk level at 1×10^{-5} rather than at a less protective level (e.g., 1×10^{-4}) is intended to protect the public from an unacceptable total exposure to carcinogens.

Conservative assumptions were used to derive an upper limit estimate of risk to the various populations considered in this assessment. The calculated concentrations of heavy metals in the air and soil were compared to specific EPA criteria. Potential health risks from the inhalation of particulates were characterized by comparing modeled annual air concentrations of particulates (at the maximum point of impact) to national ambient air quality standards. Potential impacts to environmental receptors were evaluated by comparing modeled surface water concentrations to ambient water quality criteria.

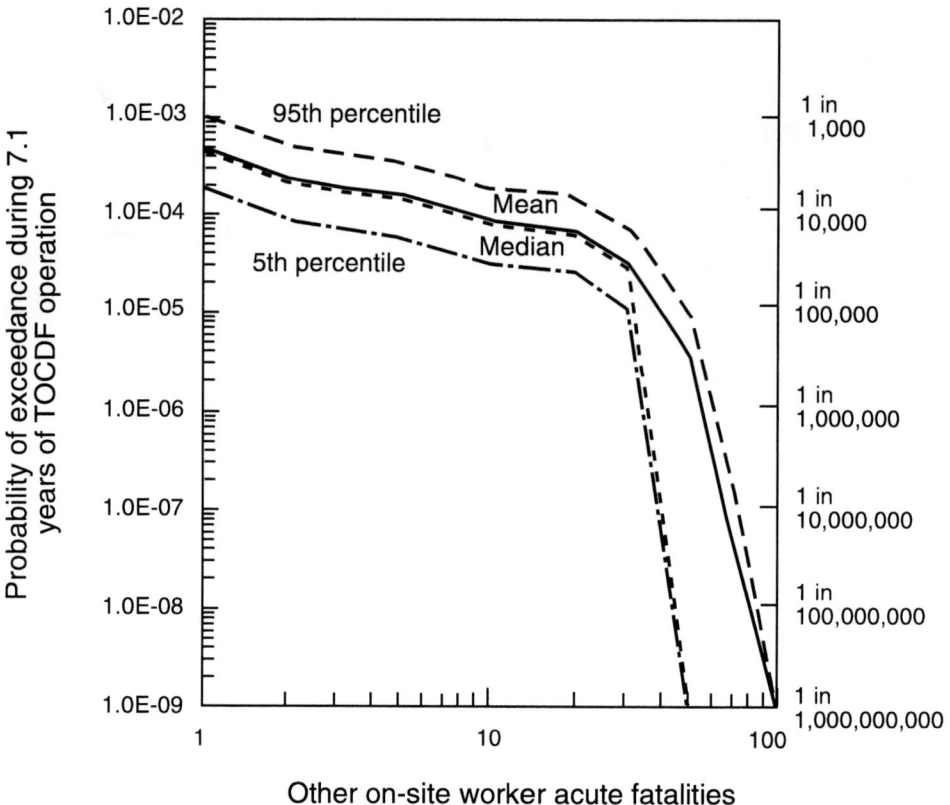

FIGURE 2-15 Acute fatalities for other on-site workers at TOCDF from accidents during disposal processing. Source: Adapted from U.S. Army, 1996c.

In addition to the conservative assumptions for selecting parameters for the HRA, a number of scale-up adjustments were made to the data used to model air and deposition concentrations. The HRA was conducted using data from JACADS to represent TOCDF emissions because actual data on emissions from the TOCDF incinerators had not yet been collected. The JACADS emissions data were scaled up to reflect the maximum anticipated feed rate at the TOCDF and modified to reflect the maximum metals composition in munitions to be treated at the TOCDF.

The emissions data were also modified to reflect potential upset conditions when the incinerator units might emit higher than usual concentrations of constituents. As suggested in the EPA screening assessment guidance (EPA, 1994), it was assumed that 5 percent of the time organic emissions from the incinerators would be 10 times higher than normal and that 20 percent of the time metals emissions would be 10 times higher than normal. These assumptions were chosen to account for abnormal combustion conditions that might occur during startup, shutdown, or production upsets, and are very conservatively based.

To account for unknown organic constituents (i.e., constituents that were not specifically analyzed by the laboratory analytical methods), the estimated emissions were weighted according to a recommended EPA method (EPA, 1994). This adjustment was made only for data for which associated total organic carbon information was available for the emissions (the method requires total organic carbon data). This approach assumes that the unidentified organic compounds are similar in toxicity and chemical properties to the identified organic compounds taken as a whole. Computations are made by increasing the emission rate of each identified organic compound by the ratio of the concentration of total organic compounds to the total concentration of all identified organic compounds. The risk assessment was then conducted using the adjusted (i.e., increased) emission rates for each identified organic compound. Total organic carbon data were available for the metal parts furnace, the dunnage incinerator, and the liquid incinerator units (Utah DSHW, 1996). The adjustments to the JACADS emissions data increased the overall risks associated with operation of the TOCDF and provided a conservative screening risk assessment (EPA, 1994).

HRAs were conducted for an adult resident, a child resident, a subsistence fisher, and three different farmers, an approach that is consistent with EPA guidelines. The adult resident was assumed to reside for 30 years at the maximum off-site point of impact along the northern boundary of the TOCDF facility. The child resident was assumed to reside for six years at the same maximum point of impact. The subsistence fisher was assumed to reside for 30 years 40 kilometers (25 miles) north-northwest of the TOCDF, where subsistence fishing was thought to be practiced. The three farmers were assumed to reside for 40 years in the vicinity of the TOCDF.

Pathways for human exposure to incinerator vapor and particulate emissions were specified in the HRA for each of the six individuals analyzed. Exposure pathways included: consumption of fish, meat, and homegrown vegetables; incidental ingestion of soil; and inhalation. For example, in the case of the subsistence fisher, the fish consumption rates used in the analysis were considered to be representative of a subsistence fisher rather than the general population; all fish consumed were assumed to be caught in water impacted by incinerator emissions; and 25 percent of aboveground and below-ground vegetables consumed was considered to be homegrown. In the case of Farmers A and C, 100 percent of the beef was assumed to be from home-raised stock that grazed in various locations near the TOCDF. For Farmer B, 12.5 percent of aboveground and 100 percent of below-ground vegetables consumed were assumed to be homegrown and contaminated by emissions from the TOCDF.

The study assumed that all individuals were exposed 350 days per year, except Farmer A, who was assumed to be exposed 175 days per year. For all individuals, the TOCDF was considered to be operating continuously (i.e., 24 hours per day, 365 days per year) for a period of 10, 15, and 30 years. Incinerator emissions evaluated in the risk assessment included:

- each incinerator or heating, ventilation, and air conditioning filter stack operating individually
- the combined stack (liquid incinerators, metal parts furnace, and deactivation furnace system operating simultaneously)
- maximum TOCDF operations (all TOCDF units operating simultaneously)
- maximum TOCDF plus CAMDS operations (all TOCDF units plus CAMDS deactivation furnace and heating, ventilation, and air conditioning filter stack operating simultaneously). The latter scenario, involving total TOCDF operations plus CAMDS operations, resulted in the largest overall cancer risk.

TABLE 2-1 Summary of the Human Health Risk—Overall Risk of Cancer for Combined TOCDF and CAMDS Disposal Operations

Receptor	Period of Operation		
	10 years	15 years	30 years
Adult Resident	$< 1 \times 10^{-6}$	$< 2 \times 10^{-6}$	$< 4 \times 10^{-6}$
Child Resident	$< 3 \times 10^{-6}$	$< 3 \times 10^{-6}$	$< 3 \times 10^{-6}$
Fisher	$< 5 \times 10^{-8}$	$< 5 \times 10^{-8}$	$< 7 \times 10^{-8}$
Farmer A	$< 8 \times 10^{-6}$	$< 8 \times 10^{-6}$	$< 8 \times 10^{-6}$
Farmer B	$< 1 \times 10^{-7}$	$< 1 \times 10^{-7}$	$< 2 \times 10^{-7}$
Farmer C	$< 9 \times 10^{-6}$	$< 1 \times 10^{-5}$	$< 1 \times 10^{-5}$

Source: Adapted from Utah DSHW, 1996.

The summary of the human health risk calculations are shown in Table 2-1. In no case did the carcinogenic risks exceed the 1×10^{-5} carcinogenic risk level established in the EPA screening risk assessment guidelines (EPA, 1994).

KEEPING ASSESSMENTS CURRENT

Quantitative Risk Assessment

The results presented in this report reflect the results of the QRA as of December 1996. The committee saw several earlier drafts of the QRA, and the risk estimates were different, sometimes significantly, in each draft. For example, in the first draft, it became obvious that several risk reduction options were available, which were highlighted in the *Systemization* report (NRC, 1996b). They included deferring the processing of wet-eye bombs until some safety concerns had been addressed, which resulted in a risk-based reordering of the disposal schedule. Operation of the liquid propane tank was also modified to reduce flammable fuel inventory in the event of seismic damage. As changes were implemented, the QRA was revised accordingly.

Because the QRA process was interactive and included input and comments from the Expert Panel and data from accumulating experience captured in the Army's "Lessons Learned" program and other site-specific risk management and reviews, more changes were incorporated. As the QRA proceeded, some major sources of risk were analyzed more thoroughly than they had been in the first draft. For example, outside experts were brought in to refine the estimates of the seismic response of structures and stacked munitions, which were subjected to a much more detailed analysis. Some of the updated analyses have led to significant reductions in the estimates of stockpile risk and to some reductions in uncertainty.

As this discussion shows, QRAs represent a state-of-knowledge and a state-of-the-system at a given point in time. Because both of these factors change with time, the Army is treating the QRA as a "living model" that will continue to be updated as changes affect risk or new information becomes available. This "living model" QRA will be the basis for ongoing risk management at the TOCDF. For example, an analysis of stockpile failure modes under seismic conditions might lead to ideas for further reducing interim stockpile risks.

Health Risk Assessment

The HRA reflects pre-operational assumptions about emissions from the incinerators and the frequency of upset conditions. As the TOCDF continues operation and actual operational and trial burn data become available, these levels will be substituted for the initial assumptions.

ANALYZING AND INTEGRATING RESULTS

Different types of risk factors are evaluated on different bases, making the integration of results difficult. However, some comparisons can be helpful. The QRA for Tooele estimates the risk to an individual living within a ring 2 to 5 kilometers (1.2 to 3.1 miles) from the TOCDF. This mean risk level can be interpreted as typical of the individual public fatality risk from an agent release because about half of the people in the

---

BOX 2-1
Individual Risk at DCD and the TOCDF in Perspective

According to the QRA, the public mean individual fatality risk levels in the 2 to 5 kilometer zone are as follows:

- during one year of exposure to the storage area, 6.4×10^{-6} (1 in 160,000)
- during one year of exposure to processing operations, 1.7×10^{-7} (1 in 6 million)

The disposal-worker mean individual fatality risk during one year of exposure to processing operations is: 4×10^{-5} (1 in 25,000).

To put these numbers in perspective, consider that the risk for a typical American of dying from a fall is about 7×10^{-5} per year (1 in 14,000), and the risk of dying in a fire is about 3×10^{-5} per year (1 in 33,000). The risk of dying from a lightning strike is about 5×10^{-7} per year (1 in 2 million).

---

zone will have a somewhat higher risk and half will have a somewhat lower risk. Note that government property around the TOCDF provides at least a 2-km buffer zone. Individual risk levels decrease with distance from the facility. These results are summarized in Box 2-1.

An individual living in the 2 to 5 kilometer ring around the TOCDF is subject to the four general categories of risk shown in Table 2-2. The numbers shown are the probabilities that that individual will die or will contract cancer as a result of either the 7.1 year TOCDF processing schedule or of the same 7.1 year period of exposure to storage of the full stockpile. In fact, once disposal has begun, an individual would be exposed to both the processing risk and a diminishing stockpile risk as the stockpile was depleted (Figure 2-8). For an individual who lives more than 10 kilometers away, the individual risk of fatality associated with disposal accidents is insignificant compared to his or her exposure to ordinary risks, and the residual risk of fatality of 4×10^{-8} per year is from stockpile storage accidents.

The HRA only estimates the risk to a "maximally exposed individual" and does not address health risks to more distant individuals because atmospheric dispersion dilutes source streams with distance. This is

TABLE 2-2 Risks for an Individual Living 2 to 5 Kilometers from the TOCDF

Consequence	7.1 Years of Exposure to DCD Storage during Disposal Processing	7.1 Years of Exposure to Disposal Processing at the TOCDF
Disposal on Schedule		
Acute Fatality from Agent Release Accidents	4.5×10^{-5} probability of fatality	1.2×10^{-6} probability of fatality
Latent Fatality from Agent Release Accidents	8.5×10^{-10} probability of delayed cancer (only HD is a carcinogen)	1.7×10^{-11} probability of delayed cancer (only HD is a carcinogen)
Eventual Chance of Cancer from Exposure to Stack Emissions	Not applicable	Less than 1×10^{-5} probability of delayed cancer to a "maximally exposed individual"
Delayed Disposal		
	Total storage risk increases linearly with time	Disposal risk shifts to the future

Source: Adapted from U.S. Army, 1996c; Utah DSHW, 1996.

the same reason that minor stockpile accidents have no impact beyond 10 kilometers.

There is a 1 in 22,000 (4.6×10^{-5}) chance of a fatality to a person located 2 to 5 kilometers from the TOCDF from stockpile or storage accidents over a 7.1 year period (Table 2-2). The health risks from normal operations are computed in the HRA as a conservative upper limit for a maximally exposed individual outside the site. The criterion established by the EPA of 1×10^{-5} probability of delayed cancer is based on a very conservative estimate of the risk. With proper treatment, not all of these cancers will cause death. All that can be concluded is that the cancer risk from normal operations is lower, probably much lower, than the fatality risk from accidents.

The EPA, which has set the basic standards and approach for the HRA, suggests an acceptable risk of a lifetime probability of getting cancer of less than 1 in 100,000 for a 70-year lifetime after exposure to the health risk associated with the facility. For an average one year period within that person's lifetime, the risk of latent cancer would be roughly on the order of 1 in 7 million. To put these risk criteria in perspective, consider the familiar cancer risks, some of which are voluntary (lung cancer is linked to smoking), others of which are not. The general incidence of lung cancer in the U.S. population means that the annual chance of an individual dying are about 1 in 2,000, an annual risk level of 5×10^{-4}. For all cancers, the chance is about 1 in 600 per year. (Note that there are about 500,000 deaths from all cancers and 150,000 deaths from lung cancer each year in the United States.) Thus, the change in cancer risk as the result of exposure at the maximum level considered acceptable by the EPA would be much less than 1 percent of the normal incidence of all cancer.

Table 2-3 presents measures of public risk (expressed as the number of fatalities expected as the

TABLE 2-3 Expected Number of Fatalities (Societal Risk)

	Processing Period	
Consequence	7.1 Years of Exposure to DCD Storage during Disposal Processing	7.1 Years of Exposure to Disposal Processing at the TOCDF
Fatality from Agent Release Accidents	Public risk, 1.6×10^{-3} expected fatalities 2.4×10^{-6} risk of latent cancers Worker risk not estimated in QRA	Public risk, 1.6×10^{-4} expected fatalities 2×10^{-8} expected cancers Disposal workers, 1.3×10^{-1} expected fatalities Other workers, 6.6×10^{-4} expected fatalities
Eventual Chance of Cancer as a Result of Exposure to Stack Emissions	Not applicable	Not included in HRA

Source: Adapted from U.S. Army, 1996c.

result of storage and disposal operations). Worker risks are also shown. The HRA does not address public risk; thus the health risk from normal operations cannot be put into the same framework. The HRA sets a criterion for a "maximally-exposed person" but does not evaluate the number of individuals who might be exposed at that or lower levels. The criterion is set low enough to compensate for the potential of multiple exposures. The EPA has established acceptable risk criteria for the potential for latent cancers associated with plant operating emissions. However, criteria have not been established by a regulatory group for risks associated with accidents.

The foregoing discussion of DCD/TOCDF risk assumes the timely disposal of agent over a 7.1 year period. Delays in processing would have a minimal effect on the disposal risk, but the risk of continued storage would increase.

3

Risk Management in the Chemical Stockpile Disposal Program

The principles of risk management are presented in the last section of Appendix A, where risk management is described as the process by which risks are understood and controlled. Step 2 of the risk management process at the end of Appendix A also provides some general examples of different options for risk reduction that can be used in risk management at DCD/TOCDF. All affected parties have roles to play in the risk management process. The Army is responsible for managing the chemical stockpile and its destruction. However, the Army's contractors, individual workers, local governments, and the affected public must all participate for the process to proceed efficiently and safely (NRC, 1996c). Risk management usually involves the following steps:

- understanding the risk (including identifying major contributors to risk)
- suggesting alternative ways to reduce risk
- evaluating risk reduction alternatives
- selecting preferred alternatives (including implementing decisions)

REQUIREMENTS FOR RISK MANAGEMENT AT DCD/TOCDF

As techniques for risk management have changed rapidly over the past 15 years, approaches to risk management at DCD/TOCDF have also evolved. In the early history of the stockpile, surveillance and maintenance were internal Army responsibilities. When the TOCDF was planned and constructed, the Army focus expanded to comply with regulations imposed by outside authorities with jurisdiction over disposal operations. Stemming from a growing base of knowledge and experience in risk management, and with the encouragement of the Stockpile Committee (NRC, 1996b), the Army is now attempting to move beyond a command and control compliance culture toward a more open and comprehensive risk management process.

The types of risk-related activities that must be managed at DCD/TOCDF are described in Chapter 2 and include the storage and handling of agent and munitions and the operation of the agent destruction facility. The QRA and HRA studies provide a current indication of the sources of risk, the magnitude and distribution of risks, and the levels of uncertainty. Parties interested in DCD/TOCDF operations include the Army, Army contractors, site workers, the local community, communities near other stockpile sites, state and local governments, state emergency preparedness programs, and state citizens advisory commissions (CACs). For risk management to be effective, each group needs to understand the assessment process, the results, and the significance of the results. This requires:

- effective risk communications among the various levels of programmatic and on-site Army and contractor personnel
- effective risk communications to technical audiences
  - presentation of materials in language understandable to engineers, technicians, and technical managers
  - presentation of detailed material for QRA/HRA practitioners and risk managers, including emergency preparedness officials and local regulatory groups

- establishment of avenues of communication with plant managers, technical staff, contractors, and regulators to facilitate cooperative interaction on matters of risk management and policy
• effective risk communications to nontechnical audiences, which are essential to the community involvement program
  - provision of reports and presentations to the CAC and the public
  - establishment of avenues for the community to let the Army know how presentations could be more useful and what additional risks and concerns they would like to see analyzed
  - regular interaction on matters of risk management policy

Finally, effective risk management requires a program to track the present status of risk estimates, monitor changes that might shift risk levels significantly, evaluate suggestions for risk reduction, and assess proposed operational improvements in a way that considers the concerns of interested parties.

EVOLUTION OF THE RISK MANAGEMENT PROGRAM

Current Status

The current CSDP risk management program is a multilevel program that defines policy, sets requirements, provides guidance on implementation, and, at the facility level, defines specific requirements the facility must meet and specific management processes that must be implemented. The CSDP risk management program is evolving and has been formalized in the last two years, based on a long history of safety and hazard analysis and regulation by the Army. An informal risk management process was developed at the TOCDF parallel with the site-specific QRA. This process was described in the Stockpile Committee's March 1996 report, *Review of Systemization of the Tooele Chemical Agent Disposal Facility* (NRC, 1996b), which summarized a number of changes that had been implemented as a result of accident scenarios identified in preliminary work on the QRA. As part of the risk management process, the following risk monitoring activities have been introduced:

• performance evaluation (based on feedback from activities and incidents)
• emergency response exercises (periodic exercises on site, with CSEPP personnel)
• risk tracking (as new data become available, as risk models are improved, and when changes occur in the facility, the related changes in risk related to safety, environmental protection, and emergency preparedness will be calculated and tracked)
• lessons learned programs (PMCD now invites all facilities to participate in meetings about design lessons learned and programmatic lessons learned)

The committee supported these activities and recommended that they be formalized before the end of the first year of agent operations. However, the Stockpile Committee is concerned that the emphasis on safety at the TOCDF has been focused on agent-related issues, with a corresponding lack of emphasis on traditional industrial safety practices and procedures (NRC, 1996b). The committee believes that failure to wear required protective equipment (e.g., to protect the eyes), poor housekeeping, and unsafe conditions (e.g., obstructions blocking access to safety equipment and walkways), which were observed by the committee during recent site visits to the TOCDF, indicate the lack of an established safety culture or mind-set. Although the absence of a pervasive safety culture that emphasizes agent-related and nonagent-related safety matters equally is not likely to change the QRA public risk estimates, it may significantly increase worker risk.

The first step in the Army's attempt to formalize the risk management process was the publication of the *Tooele Chemical Agent Disposal Facility Risk Management Plan* in April 1995 (U.S. Army, 1995b). After several reviews, that document was replaced by a program-wide document, *Chemical Agent Disposal Facility Risk Management Program Requirements* (U.S. Army, 1996e), which provides a basis for the CSDP risk management program. The risk management program is a framework for understanding and controlling all elements of risk within the disposal facility and the stockpile storage area. It links risk management needs to other specific requirements of the Army and other parties at top levels of management and identifies specific documents and references that apply to all CSDP facilities (as recommended in the committee's *Systemization* report [NRC, 1996b]).

Currently, two additional steps are being taken—plans for a site-specific risk management program and a programmatic policy guide are being developed. *The TOCDF Risk Management Program Plan* (EG&G, 1996), initiated by Edgerton, Germerhausen and Grier, Incorporated (EG&G), the site contractor at the TOCDF, includes definitions of contractor responsibilities as well as responsibilities of various risk management program elements. The plan also includes a "Compliance Matrix" that identifies site implementing documents used to meet the requirements listed in the program requirements document cited above (U.S. Army, 1996e). The site-specific plans are to be "living documents" (i.e., they will be continually updated.)

In January 1997, the Army issued its draft, *A Guide to Risk Management Policy and Activities* (the *Guide*) (U.S. Army, 1997c). This draft provides an overview of the processes for managing risks associated with PMCD activities. It defines risk terminology and describes risk management processes and decision bases. The role of each organizational element is described in terms of specific risk management activities: assessment, requirements, monitoring, management of change, and public participation. The rest of the draft *Guide* explains these activities in some detail, identifying the products of each organizational element, the evaluation tools, the source of authorization, and the techniques for tracking performance. This draft breaks new ground in Chapter 7 by presenting a process for managing changes that may affect the risk associated with PMCD activities. It defines issues that are matters of risk assessment and issues that are matters involving policy (value judgments) and attempts to establish an approach to integrating them and involving the public in that integration.

The draft *Guide* explains the PMCD's risk management policy in the context of the organizational structure shown in Figure 3-1 (p. 63 of the *Guide*). Although the role of public affairs programs and two-way communications between the Army and the public are acknowledged, and a specific public participation element is defined for the change management process, Figure 3-1 suggests that the public outreach program is separate and will not be integrated into the risk management process in a manner consistent with past NRC recommendations (NRC, 1996b, 1996c).

The PMCD policy indicates that risk management is integrated into the normal functioning of the organization:

- Operations are now based on the Risk Management Program Requirements document (U.S. Army, 1996c).
- The Risk Management and Quality Assurance Office has been assigned the task of integrating risk management for operations, design, and construction.
- The Environmental and Monitoring Office has been assigned the task of assessing hazards to the environment, the populace, and biota in terms of regulatory requirements.
- The CSEPP has been assigned the task of planning for potential emergencies and providing liaisons with other emergency preparedness organizations.
- The Public Affairs Office is charged with providing liaisons among the public, the CAC, state authorities, and the Army to facilitate public involvement.

Another significant element in risk management is the management of change. Although changes are usually made for good reasons, overall safety of the facility could be compromised if the effects of change on risk levels are not understood. Changes need to be documented and analyzed to see if they affect procedures, training, or other aspects of the program. In Figure 3-2 (p. 46 in the *Guide*), the change process is initiated from an "Established Configuration," which is the current state of the facility. This Established Configuration is based on the initial design of the facility and incorporates changes that have been approved and implemented. The Established Configuration is the basis for the plant's up-to-date HRA and QRA.

A description of how TOCDF evolved to the Established Configuration can be found in Example 1 later in this chapter. The change process continues with the definition of a change package and a coarse screening process to preclude detailed analyses of changes that do not significantly affect risk. The effects of relevant changes on risk are then determined by revising the

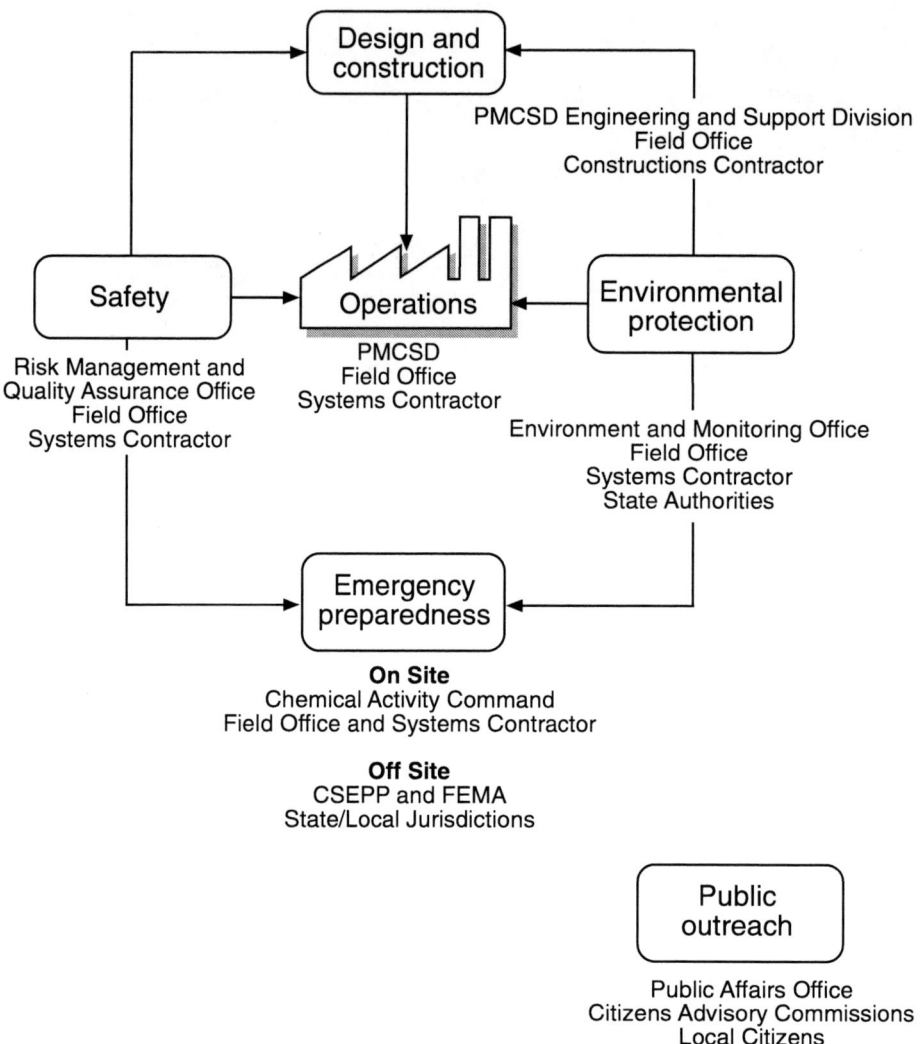

FIGURE 3-1 PMCD's organizational elements directly related to risk management (p. 63 in the *Guide*). Source: U.S. Army, 1997c. Note: FEMA is the Federal Emergency Management Agency.

HRA, QRA, and other Army safety evaluations. If a proposed change meets HRA standards and its effects on the QRA, the disposal schedule, and costs are documented, then the proposed change is presented for public comment.

If a change is significant, assessing its value is acknowledged to be both a policy question and a factual question. Structured discussions focus attention on all factors that affect the decision. Information on the impact of the proposed change is made available to the public, the CAC, and state regulators, and public comments are solicited. For the most significant changes (RCRA Class 3 modifications), the Army will conduct a public workshop. The Army's decision will take into account community desires and needs as well as important facts and intangible factors, which are summarized in Table 3-1 (p. 53 in the *Guide*). Note that factor 6 in Table 3-1, "comparison to previous decisions," ensures either that decisions are consistent or that the reasons for inconsistencies are clearly stated. A thorough consideration of uncertainties is also required. The Army will prepare responses to all public comments and inform regulators and the CAC of their decisions and rationale.

The committee agrees with the Army that "each change proposal is likely to involve unique circumstances and factors, so it is not possible (or desirable) to prescribe a set decision process with fixed criteria."

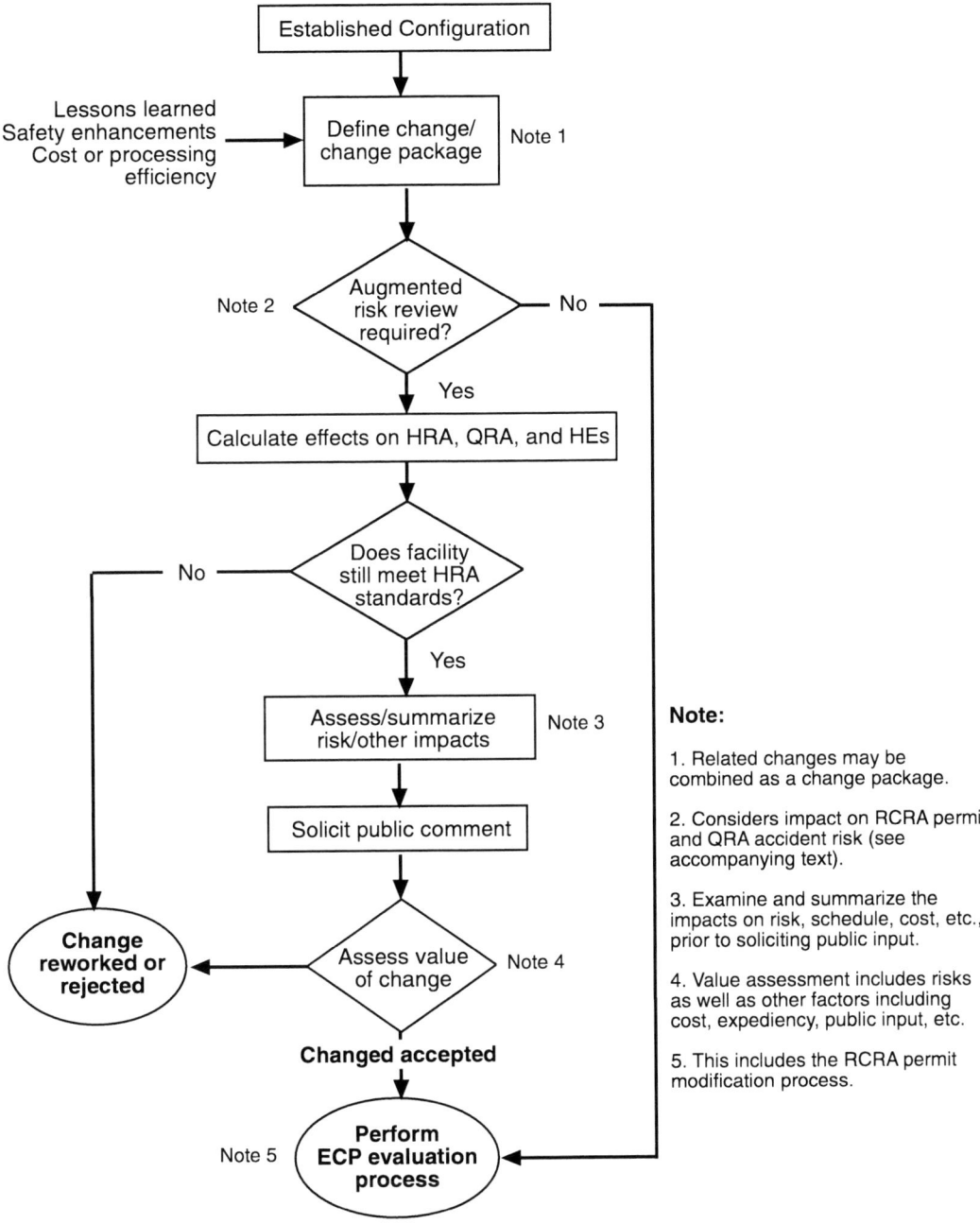

FIGURE 3-2 The change process (p. 46 in the *Guide*). Source: U.S. Army, 1997c. Note: ECP means engineering change proposal; HE means hazard evaluation.

If all issues are considered in an appropriate and timely manner, general consensus may be possible. But even if consensus is not reached, the Army, as decision maker, will provide a "synopsis of the considerations and a summary of the overall decision basis, listing the rationale for each factor." In this way, interested parties can see if their concerns were considered and the affect they had on the decision.

In general, the Stockpile Committee agrees with the proposed management of change process developed by the CSDP and encourages its use. As this process is applied to change proposals, the Army will learn a great deal about the utility, benefits, and difficulties of the process, which may lead to improvements. The committee is concerned that the proposed process will not work well unless the current description of public

TABLE 3-1 Issues and Factors in Assessing the Value of Change Options (p. 53 in the *Guide*).

1. Public Input
2. QRA Risk
   a. All available QRA risk measures, including expected fatalities, cancer incidence, fatalities at a one-in-a-billion probability, and probability of one or more fatalities
   b. Risk tradeoffs: public versus worker, individual versus societal, processing versus storage
   c. Uncertainties in the technical assessment of risk
   d. Insights from sensitivity studies
3. Hazard Evaluations
4. HRA Risk
   a. Insights from sensitivity studies
5. Programmatic
   a. Cost of the change relative to other proposals and program objectives
   b. Schedule for implementation
   c. Uncertainties in estimates
   d. Impact of implementation on overall objectives and schedule for disposal of the weapons and chemical agent
   e. Consideration of the improvement anticipated by this change with other proposed improvements
6. Comparison to previous decisions

Source: U.S. Army, 1997c.

outreach is expanded to include public involvement that is fully integrated into the management of change process (NRC, 1996b, 1996c). In some parts of the draft *Guide*, broader public involvement is acknowledged, but more effort will be required to develop a fully integrated approach.

Further Development of the Draft Risk Management Policy Guide

Although the Stockpile Committee generally agrees with the January 1997 draft, *A Guide to Risk Management Policy and Activities* (U.S. Army, 1997c), further development is needed in some areas. For example, the QRA has played a significant role in defining the current Established Configuration for the TOCDF, and it is reasonable to expect that the same will be true at other sites. The TOCDF is a good example of how information from the QRA can be used to facilitate risk reduction. However, Chapter 5 of the draft *Guide* barely mentions this.

The draft *Guide* only notes that interfaces and communication among the program elements are needed. It does not describe how interfaces will be managed across the functions of systemization/operation, safety, environmental protection, emergency involvement, and public participation. Some important cross-functional risk management issues are not described in detail and could benefit from a detailed description of managerial responsibilities. The draft *Guide* does indicate that maintaining the Established Configuration is currently the responsibility of the PMCD; that updating the QRA and HRA is the responsibility of the Army's Risk Management and Quality Assurance Office; and that significant changes that impact risk need to be closely coordinated with emergency preparedness personnel, environmental managers, and the public. However, the coordination and chain of management responsibilities are not clearly defined.

The draft *Guide* appears to be an ideal place for the PMCD to elucidate the policy on safety culture; on the importance of industrial safety practices (a responsibility of each individual that can only be realized with a corporate commitment); on relationships among Army organizations, both at the PMCD level and on site; and with contractors and other agencies concerned with overall safety.

Table 3-2 (p. 14 in the *Guide*) provides a matrix of functions of the risk management plan and the five necessary activities defined in the draft *Guide*. The table shows few discrete links between the public outreach program and other activities. The committee is concerned that public outreach has not been conceptually linked to assessment, requirements, and monitoring. The public outreach function requires further thought and development.

In Table 3-3 (p. 67 in the *Guide*), the need for development of more fully-integrated public outreach and public involvement is indicated by the lack of an explicit method for tracking how public involvement and the CSEPP affect risk assessment and vice versa.

APPLYING RISK ASSESSMENT RESULTS TO RISK MANAGEMENT

Example 1: The TOCDF Established Configuration

The process described above for risk management at the TOCDF is practical and has already been tested. The Established Configuration referenced in the *Guide*

TABLE 3-2 Activities by Risk Management Function (p. 14 in the *Guide*).

Functions	Activities				
	Assessment	Requirements	Monitoring	Management of Change	Public Participation
Design & Construction	X	X		X	
Systemization & Operations	X	X	X	X	
Safety	X	X	X	X	
Environmental Protection	X	X	X	X	
Emergency Preparedness	X	X	X	X	
Public Outreach				X	X

1. *Assessment*—identifying, evaluating, and understanding hazards and risks.
2. *Requirements*—establishing criteria for safety, environmental protection, and emergency preparedness.
3. *Monitoring*—the regular trending and tracking of performance.
4. *Management of Change*—the evaluation of strategic or necessary changes to the facility or its operation in accordance with the requirements of risk management. This includes an authorization process; i.e., sign-off of the approval/acceptance chain of command needed to initiate change following its evaluation.
5. *Public Participation*—the communication of facility and operations risks and the manner in which the Army is managing those risks, the gathering of input and feedback from the public, and the use of that information in decision making.

Source: Adapted from U.S. Army, 1997c.

represents the current state of the TOCDF and is the basis for the current QRA and HRA; improvements in TOCDF safety are ongoing.

Some initial changes to reduce risk were introduced in the committee's March 1996 *Systemization* report (NRC, 1996b). First, the QRA identified potentially high risk scenarios. Then, an engineering change analysis and test options were developed. In some cases, determining the technical merits of alternatives was expected to take significant time, so the order of the processing of munitions was changed, which had no impact on the HRA but greatly reduced the QRA risk for certain scenarios based on broad uncertainties. Reordering reduced the risk, and engineering changes were made to head off potential problems. Risk management improvements to reduce disposal-related worker risk for the first two disposal campaigns were modeled in the QRA. As additional measures are implemented for succeeding campaigns, the Army expects further reductions in worker risk.

TABLE 3-3 PMCD Risk Management through Its Organizations and Functions (p. 67 in the *Guide*).

Organization/Risk Function	Risk Management Tasks		
	Evaluation	Authorization	Tracking
PMCSD/Operation	RMPR/COR	PMCSD Mission	RMPR Award Fee
RM&QA Safety	RA	RAC Matrix	RMPR/HTL Award Fee
E&M/Environmental Protection	RA	State and Local Regulations	RMPR Deficiencies Award Fee
CSEPP/Emergency Preparedness	Drills	Graded Drills	n/a
PAO/Public Participation	Feedback	Acceptance	n/a

Notes:
    Award fee = part of the contractual arrangement with CDF contractors that includes a performance-based fee, based heavily on safety and risk
    COR = Field Office Contracting Officer's Representative
    Drills = emergency response drills, feedback - information obtained from the local community
    HTL = hazard tracking log
    RA = risk assessment
    RMPR = *Risk Management Program Requirements* document

Notes added by the committee:
    E&M = Environmental and Monitoring
    PAO = Public Affairs Office
    PMCSD = Project Manager for Chemical Stockpile Disposal
    RAC = Risk Assessment Code

Source: U.S. Army, 1997c.

Changes at TOCDF Based on the QRA

Examples of how the QRA was used to make significant changes in the facility are given in the following paragraphs, which are summarized from the TOCDF QRA. The following changes are reflected in the current QRA:

Metal Parts Furnace Airlock. Earlier risk models identified the potential of a buildup and ignition of agent vapors in the feed airlock of the metal parts furnace (MPF). The most significant risk was for bulk agent containers held in the airlock for a longer-than-normal interval, but there was also a risk during the processing of projectiles. The public risk of a potential accident was estimated to be relatively small; however, the risk to workers was estimated to be higher than for other accidents. In light of these findings, the PMCD took steps to minimize the potential risk by changing the hardware to vent the airlock and minimize agent buildup. Operational changes were also made to limit the time an item could be held in the feed airlock.

MPF Processing of Weteye Bombs. The QRA postulated that the occurrence of an energetic reaction between molten aluminum and liquid agent could not be ruled out during the processing of aluminum weteye bombs. Calculations supporting the QRA indicated that molten aluminum could be present when significant agent was still left in the weteye bomb. The potential for explosive interactions of molten aluminum with water is known, although the exact conditions for explosions have not been determined. The interaction of molten aluminum and liquid agent has not been studied, but the potential for an energetic explosive reaction could not be ruled out. The QRA results indicated that this scenario could be a significant contributor to TOCDF worker risk, but the frequency and consequences were considered to be very uncertain.

Because of these uncertainties, the PMCD determined that it would be advantageous to change the TOCDF processing campaigns to exclude the weteye bombs from the first campaign. This change was made for two reasons. First, the processing campaign cannot begin without a trial burn protocol and schedule in place, and the MPF trial burn requirements for the first campaign would require larger-than-normal heel quantities (agent remaining after the draining operation).

The QRA identified the presence of liquid agent as a key element of a potential accident, and the trial burn requirement would increase the likelihood of liquid agent availability. Second, the uncertainties in the QRA results suggested that the issue required further study; a delay in processing weteye bombs allowed time for a thorough review of the calculations and the development of a strategy to eliminate the potential for aluminum-agent interaction. This work is ongoing and is to be incorporated into a QRA update before the weteye campaign begins.

Seismic Anchorage of the Liquid Propane Gas Tank. One contributor to the risk from earthquakes identified in the risk model was the 50,000-gallon liquid propane gas (LPG) tank installed at the site to provide a temporary fuel supply in case the natural gas supply via pipeline was interrupted. The QRA assessed the risk from earthquakes that exceeded the design basis and found that the LPG tank, although appropriately designed for seismic zone three, had a lower seismic fragility than other plant equipment and structures. Scenarios involving explosions and fires after an earthquake that could dislodge the tank from the outlet pipe or cause the tank to fail completely were found to contribute about one-half of the seismic risk. After reviewing this finding, the PMCD reconsidered the need for LPG and concluded that the original criterion for the size of the tank, which involved maintaining the operational status of furnaces and incinerators, was not necessary and that the LPG would only be used to provide fuel for the boilers. Thus, the fuel inventory in the tank could be reduced to less than 10,000 gallons.

The reduction had a twofold effect. First, the lower fuel volume and resultant lower weight have increased the seismic capacity of the LPG tank to the point that only much larger accelerations can cause the anchorage to fail and cause a gas leak. Second, even if the tank fails, the lower fuel quantity reduces the likelihood of an LPG explosion that would be significant enough to cause an agent release.

Campaign Order. The initial QRA results indicated that the original order of disposal did not eliminate the munitions with the highest risk first. For example, two early campaigns to destroy mustard would have had a minimal effect on the storage risk. Sensitivity studies were performed to optimize the schedule to eliminate the higher storage risk items first. The change has resulted in a significant reduction in the public risk of storage during disposal.

Stockpile Earthquake Preparedness (Mitigation). DCD has been provided with a prioritized list for inspecting igloos after an earthquake. The QRA models, augmented by detailed analyses of seismic vulnerability, were used to determine which munitions and igloos are most vulnerable to damage so the personnel checking the condition of the stockpile could do so in the most risk-effective manner. Detailed vulnerability analyses have also resulted in some reduction in the estimates of seismic risk for the stockpile.

MPF Restart Procedures. While investigating the potential for an MPF explosion on restart, QRA analysts identified a path through the procedure where warnings against restart might not be clear in certain situations. The restart procedure was modified based on this finding.

MPF Exit Airlock Automatic Continuous Air Monitoring System (ACAMS). The QRA indicates that the risk is sensitive to an operational step—ACAMS monitoring in the MPF exit airlock. Therefore, the risk management plan will identify ACAMS monitoring in the MPF exit airlock as a critical activity that requires special attention during operation.

Summary

While QRA models were being developed, interactions between the PMCD, the TOCDF staff, and the QRA team led to improvements in the QRA analysis as well as to refinements in facility operations. For example, the identification of potential accident-initiating events required the development of process operational diagrams, which involved a step-by-step delineation of facility operations. Interaction ensured that the QRA models were accurate; at the same time, the QRA process helped to refine operations.

The risk assessment/risk management process needs to be orderly. The first QRA results identified several events that were high contributors to risk, including the seismic failure of the LPG tank. These high contributors were addressed immediately (as described above). The changes resulted in a residual risk dominated by seismic events. However, these events involve earthquakes that exceed normal design codes, with mean accelerations of 0.2 to nearly 1.0 g (1 g equals the acceleration of gravity). The recurrence intervals for earthquakes of this severity in the Tooele area are greater than one thousand years. Since the weakness in the LPG tank was corrected, the remaining risk is reasonably low. The remaining significant contributors to seismic risk are stacks of stored munitions that could tip and fall leading to agent leakage or explosions and fire, steel-arch igloos that could collapse and crush munitions, and structural failures in the container handling building/unpack area of the processing facility.

Two risk management processes are continuing: (1) the QRA team is examining all contributors at all stockpile sites to consider mitigation measures akin to those adopted at TOCDF for any high contributors; and (2) the management of change process described earlier in this chapter is being developed to ensure that changes proposed for any reason do not inadvertently increase risk. A discussion on using carbon filters with the baseline system (Example 2 later in this chapter) is provided because a decision on using them will probably be the first full-scale application of the management of change process. The addition of carbon filters has been under consideration for some time.

The ongoing use of the QRA in risk management has been effective in optimizing facility operations and improving safety. The PMCD reviews of the QRA include examining all of the assumptions and verifying all models. The QRA allows for the continuous investigation of ways to minimize the risks of given operations, which can, and has, resulted in refinements to operations and procedures.

The changes listed in Example 1 illustrate the effectiveness of using the QRA as a tool for risk management and indicate the acceptance of the QRA as an integral part of risk-informed decision making by the PMCD and the TOCDF staff. The interaction among the PMCD, the TOCDF staff, and the QRA team, and resulting improvements, were the basis for the development of the new management of change process that includes explicit consideration of effects on the HRA and solicits public comments as input to the decision process. Although public comments were not used as input to establish the initial system configuration, they will be solicited for major changes in the future.

Example 2: Carbon Filter System for the Pollution Abatement System

The first full-scale application of the new management of change process is expected to be the evaluation of the proposed change to modify the pollution abatement system (PAS) by adding carbon filters to control emissions from the incinerators. An earlier NRC Stockpile Committee report, *Recommendations for the Disposal of Chemical Agents and Munitions* (NRC, 1994b), recommended, in part, the following:

> The application of activated charcoal filter beds to the discharge from baseline system incinerators should be evaluated in detail, including estimations of the magnitude and consequences of upsets, and site-specific estimates of benefits and risks. If warranted, in terms of site-specific advantages, such equipment should be installed.

This recommendation was prompted by the committee's belief that adding a carbon filter system downstream of the existing PAS might provide further protection against an accidental release of agent from the stack (impacting the QRA) and might further reduce volatile organic emissions during normal and upset operations (impacting the HRA), even though organic emissions have been shown to be at trace levels and below the level of regulatory concern. The committee also recognized that adding a PAS carbon filter system might have adverse effects, e.g., the potential for filter fires and the consequent sudden accidental release of contaminants stored in the filter, and increased worker exposure to dioxins/furans and agent during disposal of the carbon filters.

In response to the NRC recommendation quoted above, the Army developed a conceptual design of a PAS filter system (PFS) and conducted a preliminary generic assessment of the application of this system to the common stack gases from the PAS of the baseline incineration system furnaces at the TOCDF. This evaluation concluded that the PFS could potentially enhance environmental performance of the baseline system but would increase the cost and complexity of the system. At the time (early 1994), data were insufficient to determine whether an installation was warranted because specific filter performance data and site-specific QRAs and HRAs had not been completed. Since then, site-specific QRAs and HRAs are either nearing completion or have been completed for most of the remaining stockpile sites.

Consistent with CSDP policy outlined in the draft *Guide* (U.S. Army, 1997c), the Army has developed a draft evaluation methodology for assessing PFS risk (U.S. Army, 1996f). The methodology requires the Army to conduct site-specific evaluations of the PFS, applying the same risk assessment methods, i.e., QRA and HRA, used to evaluate the risks of the established configuration of the baseline system. The need for site-specific evaluations derives from variations in site-specific factors, such as the types of chemical agents and munitions to be processed, the proximity of population, and different meteorological conditions.

The PFS can enter the new management of change process as a proposed change to the baseline system in one of two ways:

- a way to achieve regulatory compliance if the HRA indicates that the existing system does not comply with health risk standards because of anticipated levels of pollutants in emissions
- a safety improvement of the existing baseline system configuration (i.e., by reducing risk estimated by the QRA) if the existing baseline system HRA results satisfy established health standards

The first step will be to review the site-specific HRA for the baseline system established configuration to confirm that it meets the health risk standards established for the facility. If it does, the installation of a PFS is not warranted on the basis of health risk. If the results of the site-specific HRA indicate that the facility would still be in regulatory compliance with the PFS, a sensitivity analysis of the HRA-derived risk estimates will be performed. This analysis is expected to confirm the conservatism in the current HRA methodology, thereby providing additional assurance that reductions in the calculated health risk estimates, which are already below established regulatory thresholds, will have no practical benefits. Potential benefits of the PFS will then be examined from a broader perspective, such as the impact on the QRA, cost effectiveness, or public acceptance. This proposed methodology appears to be appropriate for evaluating whether the PFS should be implemented on a site-specific basis and is consistent with the Army's new management of change process as described in the *Guide*.

Availability of Data for the PFS Evaluation

Carbon filters are not usually used for treating stack emissions from hazardous waste incinerators. The optimal design for a chemical demilitarization facility must be determined from analyses of the exhaust gases, carbon bed filter performance, and anticipated plant upsets. Designs must include data for both the benefits (filter performance) and the risks (plant upsets and a delay in the destruction of the chemical weapons stockpile).

Filter Performance

The performance of granulated activated carbon under the conditions of interest, initially determined from the published literature, had to be augmented by carefully planned and controlled laboratory experiments to ensure the correct modeling and accurate predictions of results. Laboratory tests have been completed, and a simulation model of the system has been developed (U.S. Army, 1997d). Members of the Stockpile Committee have been actively following the development of the carbon filter simulation model for the past two years. The model indicates that the carbon filters will effectively remove the hazardous components of principal interest (i.e., dioxins/furans) to levels below the detection limits (if they are present at all) for at least one year, prior to bed replacement. Detection limits are the minimum levels assumed in a normal HRA analysis.

Dioxin/furan emissions at JACADS were extremely small (NRC, 1994a). The use of natural gas at the TOCDF for supplemental fuel (in addition to the agent being burned), rather than the JP-5 fuel used at JACADS, should further reduce emissions of dioxins/furans.

Equipment Performance

In addition to the filter model, the Army has developed a conceptual design of a generic PFS. This design involves the use of a gas conditioning system to reduce the water content of the gas and adjust the exit temperature and relative humidity so that trace impurities will adsorb on the bed without being displaced by water vapor. A subsequent vertical stack of fixed horizontal carbon beds is similar to the carbon filters in the existing building ventilation system, except that the ventilation system beds are stacked vertically and require a stronger housing because the gas pressure is higher. Additional safeguards are provided for the PFS to reduce the probability of an equipment failure, which could result in hot gases reaching the filters causing the desorption of contaminants from the filters.

The installation of a PFS introduces the following possible sources of risk:

- sudden releases of accumulated hazardous contaminants (e.g., dioxins/furans) to the atmosphere at higher concentrations than are generated during normal operations with no filter
- additional plant downtime because of gas handling equipment failures and the resulting extension of the stockpile storage risk (delays in processing automatically result in increased risk from storage)
- worker exposure to accumulated hazardous contaminants during replacement and disposal of the carbon filters

If there were a sudden release of agent that had accumulated on the filter, the concentrations could be above the lower detection limit. The HRA for the baseline system without a PFS requires the assumption of agent in the stack gases at the lower detection limit, so the accumulation of agent on the PFS during normal operations must also be accounted for. Thus, potentially, agent would be available for a sudden release. This possibility must be taken into account to ensure consistency between the QRA and HRA.

In summary, the committee supports the PFS evaluation methodology and finds that the carbon filters will effectively remove dioxins/furans to levels below the detection limits, with a useful bed life of at least one year. However, more information on the effects of the PFS on QRA risks, on costs and schedules, and on public concerns must be evaluated before the Army can make a final decision.

4

Findings and Recommendations

OVERVIEW

The Stockpile Committee has reviewed the PMCD and DCD/TOCDF risk management program. Specifically, the committee has examined the TOCDF risk assessments (QRA and HRA), the DCD QRA, the QRA Expert Panel's review report, the PMCD and TOCDF risk management policy and implementation documents, and the PMCD draft evaluation of charcoal filters for the PAS (U.S. Army, 1996c, 1996d, 1996e, 1996f; 1997c, 1997d; Utah DSHW, 1996; EG&G, 1996; MITRETEK Systems, 1996).

In the 1993 letter report (NRC, 1993b), the Stockpile Committee recommended that a site-specific QRA that fully displays uncertainties be conducted using "modern up-to-date methodologies...by organizations with recognized expertise in the field." The committee further stated that "independent peer reviews are an absolute requirement" and "that local representatives of neighboring communities must be involved early" so their concerns could be considered in the analytical process.

The site-specific TOCDF QRA meets the goals stated by the committee and is much improved over the earlier draft reviewed in the *Systemization* report (NRC, 1996b). The QRA is now both more thorough and more complete, the treatment of uncertainties is more rigorous and consistent, and the mechanistic models and supporting data are more convincing. An independent peer review panel conducted in-depth reviews as the work progressed and issued many insightful comments. The QRA team responded thoroughly to all comments, which led to significant improvements.

With respect to public involvement, early attempts were made to involve state officials and the Utah CAC. However, as the committee noted in the *Systemization* report, these efforts were not very productive. Plans are being developed for early public involvement at other sites and for expanding public involvement during operational (risk management) phases at the TOCDF. The Stockpile Committee supports these efforts, encourages their effective implementation, and reiterates its position that public involvement is a PMCD management function and not strictly a public affairs office function (NRC, 1996c).

The following findings are based on the Stockpile Committee's general evaluation of the PMCD's risk management program, on knowledge and observation of the TOCDF baseline incineration system and the DCD stockpile, on information provided by the Army, the Army contractor SAIC, and others, as well as on seven site visits to the TOCDF (November 1991, March 1993, May 1994, March 1995, March 1996, December 1996, and March 1997). Four subgroups of the committee also visited the TOCDF in the spring of 1995. The committee's community involvement subcommittee held meetings with the Utah CAC and other interested parties in Tooele County in March 1995 and again with the CAC in December 1996 and March 1997. The committee's risk analysis subcommittee sent representatives on five occasions to meetings of the QRA Expert Panel, SAIC, and PMCD personnel (February 1995, March 1995, May 1995, August 1995, and September 1995). In addition, committee representatives monitored a number of telephone conferences among members of the Expert Panel and the QRA team during 1996.

FINDINGS

Risk Assessments

The Stockpile Committee has followed the DCD/TOCDF QRA project closely since its inception and has maintained oversight of the Expert Panel independent peer review process. The QRA has achieved the goals set out in the committee's 1993 letter report (NRC, 1993b) and the *Recommendations* report (NRC, 1994b). The success of the QRA was a direct result of a skilled SAIC technical team, firm support from the Army and TOCDF personnel, and frequent and positive interactions between the TOCDF field staff and the QRA team. The resulting QRA was significantly improved during the Expert Panel review. The findings of the QRA are consistent with the interim findings in the *Systemization* report (NRC, 1996b).

Members of the Expert Panel included internationally recognized experts in quantitative risk assessment, as well as experts in chemical process safety and an expert from the state of Utah in combustion and chemical engineering. The peer review was especially effective because it was interactive. The panel met or talked by telephone with the Tooele QRA team many times throughout the project. This enabled the QRA team to respond to questions and revise their approach as the analysis evolved, making effective, cost-effective corrections and changes in direction and allowing the panel to investigate analyses in greater depth than would have been possible after work had been completed.

Risk assessment experts from the Stockpile Committee observed most meetings of the Expert Panel and found the panel's review to be thorough and technically sound. The panel was equally thorough in examining theoretical issues, practical results, and the style of presentation. In addition, they searched for possible omissions.

The Stockpile Committee concurs with the Expert Panel's findings (MITRETEK Systems, 1996):

- The methodology was sound and has extended the state of the art in several areas.
- The methodology was well implemented.
- Despite some reservations concerning a few technical aspects of this QRA, the panel was reasonably satisfied that these did not affect the overall conclusions of the QRA.

The committee finds that the interactive independent review process was effective and that the Expert Panel played a significant role in ensuring that the QRA met state-of-the-art standards in all significant respects. The Expert Panel had a significant impact on key areas of the QRA:

- The treatment of uncertainty is now more clearly addressed.
- The seismic vulnerability analysis for the liquid propane gas tank has been improved.
- The model for workers donning masks after a strong motion earthquake is more realistic.
- The mechanistic modeling of munitions handling accidents is much improved.
- Significant improvements in the QRA methodology manual (EG&G, 1996) have been made.

The committee believes that the HRA performed by the Utah DSHW, which is based on many assumptions and follows EPA-mandated protocols, is appropriate at this stage of TOCDF operations because it approximates a worst case for all evaluated parameters. The greatest uncertainty in the HRA is about the magnitude and composition of actual TOCDF emissions (emissions in the HRA were based on adjusted data from JACADS). As actual TOCDF operating parameters are established and data on the nature and magnitude of actual emissions become available, they can be incorporated into the HRA. The HRA does not provide the more realistic and detailed estimation of risk sources, impacts, and distribution provided by the QRA. The HRA screens latent cancer risk to "maximally exposed" individuals, imposes an acceptability criterion (1×10^{-5} carcinogenic risk level over a 70-year lifetime), and infers that the exposure to multiple individuals at or below the screening level is acceptable.

Risk Management

The committee finds that the TOCDF risk management plan has progressed and that positive steps have been taken, e.g., the development and limited use of guidance and implementation documents. The Army's draft *Guide* on risk management provides an overview of the overall risk management program, incorporating references to subsidiary risk management documents and activities. The *Guide* defines interrelationships among Army offices, contractor offices, and public entities that are or should be involved in risk management activities (U.S. Army, 1997a). The *Guide* is a significant step by the Army toward following NRC recommendations on risk management and public involvement (NRC, 1994b; 1996b, 1996c), particularly with regard to using risk analysis in the management of change process. The *Guide*, however, has not yet been completed and finalized. The committee's comments in this report reflect the expectation that the draft *Guide* will be improved, adopted, and implemented. The committee may evaluate its implementation in the future, but at present, the draft *Guide* falls short of expectations in several respects:

- The *Guide* does not describe the contributions of risk assessment to changes that led to the current Established Configuration, against which proposed future changes will be evaluated.
- The focus of the *Guide* is primarily on agent-related safety, rather than on developing and institutionalizing a comprehensive safety program (i.e., establishing a safety culture). The *Guide* should emphasize the importance of a safety culture and of adherence to the best established safety practices used by industrial process plants. In reviews of drafts of the risk management program requirements document, the QRA Expert Panel expressed concern "that the importance of establishing and maintaining a safety culture may not be addressed in the Army standards" (MITRETEK Systems, 1996).
- The *Guide* acknowledges that more must be done to shift the focus from public information to public involvement. The role of public involvement should be extended and integrated beyond the management of change process. This will require tracking public involvement activities and public response decisions and involving the public in additional activities, such as monitoring and emergency management. This integration should be implemented expeditiously.
- The *Guide* is not specific enough about how to ensure that workers, CSEPP personnel, and the public understand the risk analyses. Improved communications are essential and need to be part of risk communication plans.
- Although the *Guide* indicates that there are functional relationships among PMCD organizations with regard to risk management, it does not detail how management roles and communications across all the Army groups, contractors, subcontractors, and other agencies involved in the program will be integrated.
- The *Guide* does not indicate plans to track CSEPP responses to changes in risk or to document public involvement and Army responses to public input regarding risk.

On the positive side, the draft *Guide* presents a framework for managing changes in the configuration of a facility or changes in operations that may significantly affect risk levels. The framework allows for public input on significant changes through a comment process, which is followed by formal feedback to the public explaining the basis for a decision. The committee finds the proposed management of change process satisfactory and encourages its use. However, there may be opportunities to expand public participation further as the Army develops a more comprehensive public involvement program.

Application of Change Policy to the PAS Carbon Filters

The committee finds the following with regard to the evaluation of adding carbon filters to the PAS:

- The proposed methodology, if well implemented, is appropriate for evaluating whether or not to install a PFS on a site-specific basis (U.S. Army, 1996d).
- The proposed methodology for the PFS evaluation is consistent with the Army's proposed management of change process, as described in the *Guide*.
- Carbon filters appear to be effective in reducing the levels of dioxins/furans to below the limits of detection and to have a useful life of at least one year. Because these levels cannot be measured, however, credit only for reducing them to the detection limit appears in the HRA.
- The QRA calculations for the PFS must account for a potential sudden release of accumulated agent (based on HRA-assumed emissions at the lower detection limit) in case of a PFS malfunction.

RECOMMENDATIONS

Risk Assessments

Recommendation 1. The Army should update both the QRA and HRA at the TOCDF whenever changes to system design or operations occur that could affect QRA or HRA calculations to ensure that estimates of risk are current and reflect changes in operating conditions and experience, assumptions, and program status (current Established Configuration). The process for updating the QRA and HRA should be included in the *Guide*.

Recommendation 2. The Army should continue the site-specific QRA and HRA processes at all PMCD sites. The development of assessments for sites other than the DCD will be greatly simplified because much of the methodology has already been established. The Army should continue to obtain interactive, independent expert reviews of all site-specific risk assessments. The Army should heed the lessons learned from development of the TOCDF QRA and should incorporate the changes recommended by the Expert Panel.

Recommendation 3. The QRA methodology manual should be updated to reflect the significant improvements that have been made.

Risk Management

Policy

Recommendation 4. The Army should expand its draft report on risk management policy, *A Guide to Risk Management Policy and Activities*, to encourage the establishment of a "safety culture" within the PMCD and its field offices and among contractors and other government agencies. The *Guide* should elucidate the Army's policy on industrial safety, including the responsibilities of individuals and managers in the field and the definitions of acceptable performance.

Recommendation 5. The Army should develop a management plan (and include it in the *Guide*) that defines the integration of management roles, responsibilities, and communications across activities by risk management functions (e.g., operations, safety, environmental protection, emergency preparedness, and public outreach).

Recommendation 6. The Army should review and expand the current draft risk management plan to include public involvement in appropriate areas beyond the management of change process.

Recommendation 7. The Army should institutionalize the management of change process developed in the *Guide*. The Army should track performance of the change process and document public involvement and public responses to decisions. The Army should use this experience to improve the change process.

Recommendation 8. The Army should expand implementation of the risk management program to ensure that workers understand the results of the risk assessments and risk management decisions. The Army should also ensure that CSEPP and other emergency preparedness officials understand the QRA and how their activities might affect risk. CSEPP activities should be tracked by the Army as part of their risk management program.

Recommendation 9. The Army should implement their risk management plans and update them whenever necessary to ensure that they reflect current practices and lessons learned.

Evaluation of the Carbon Filter Design for the Pollution Abatement System

Recommendation 10. The Army should proceed with the application of its proposed methodology for evaluating the use of PAS carbon filters on a site-specific basis. For consistency with the HRA assumptions, the QRA should take into account the possible sudden release of agent that may have accumulated on the filter at a gas concentration equal to the lower detection limit.

References

CCPS (Center for Chemical Process Safety). 1989. Guidelines for Chemical Process Quantitative Risk Analysis. New York: American Institute of Chemical Engineers.

EG&G (Edgerton, Gemerhauser and Grier, Incorporated). 1996. TOCDF Risk Management Program Plan (CDRL EG059), Rev. 0A (draft), April 3, 1996. Prepared for the Program Manager for Chemical Demilitarization and the U.S. Army Industrial Operations Command. Tooele, Utah: EG&G Defense Materials.

EPA (Environmental Protection Agency). 1994. Exposure Assessment Guidance for RCRA Hazardous Waste Combustion Facilities (draft). EPA 530/R-94/021. Washington, D.C.: EPA Office of Solid Waste.

Haskin, F.E., M. Young, C. Ding, and K. Summa. 1995. CHEM-MACCS Version 1s Model Description and User's Guide. March 1995. Albuquerque, N.M.: Sandia National Laboratories.

IEM (Innovative Emergency Management). 1993. Reference Manual: D2PC and Hazard Analysis (August 24, 1993). Springfield, Va.: U.S. Army Nuclear and Chemical Agency (USANCA).

IOM (Institute of Medicine) 1993. Veterans at Risk: The Health Effects of Mustard and Lewisite. Washington, D.C.: National Academy Press.

Leffingwell, S. 1993. Personal communication from S. Leffingwell, Centers for Disease Control, Atlanta, Georgia, to the National Research Council Committee on Alternative Chemical Demilitarization Technologies.

MITRETEK Systems. 1996. Report of the Risk Assessment Expert Panel on the Tooele Chemical Agent Disposal Facility Quantitative Risk Assessment. McLean, Va.: MITRETEK Systems.

NRC (National Research Council). 1983. Risk Assessment in the Federal Government: Managing the Process. National Research Council. Committee on the Institutional Means for Assessment of Risks to Public Health. Washington, D.C.: National Academy Press.

NRC. 1984. Disposal of Chemical Munitions and Agents. National Research Council. Committee on Demilitarizing Chemical Munitions and Agents. Washington, D.C.: National Academy Press.

NRC. 1993a. Evaluation of the Johnston Atoll Chemical Agent Disposal System Operational Verification Testing: Part I. Letter Report to the Assistant Secretary of the Army. National Research Council. Committee on Review and Evaluation of the Army Chemical Stockpile Disposal Program. Washington, D.C.: National Academy Press.

NRC. 1993b. Letter report to the Assistant Secretary of the Army to recommend specific actions to further enhance the CSDP risk management process. January 8, 1993. National Research Council. Committee on Review and Evaluation of the Army Chemical Stockpile Disposal Program. Washington, D.C.: National Research Council Board on Army Science and Technology.

NRC. 1994a. Evaluation of the Johnston Atoll Chemical Agent Disposal System Operational Verification Testing: Part II. National Research Council. Committee on Review and Evaluation of the Army Chemical Stockpile Disposal Program. Washington, D.C.: National Academy Press.

REFERENCES

NRC. 1994b. Recommendations for the Disposal of Chemical Agents and Munitions. National Research Council. Committee on Review and Evaluation of the Army Chemical Stockpile Disposal Program. Washington, D.C.: National Academy Press.

NRC. 1994c. Science and Judgment in Risk Assessment. National Research Council. Committee on Risk Assessment of Hazardous Air Pollutants. Washington, D.C.: National Academy Press.

NRC. 1996a. Review and Evaluation of Alternative Chemical Disposal Technologies. National Research Council. Panel on Review and Evaluation of Alternative Chemical Disposal Technologies. Washington, D.C.: National Academy Press.

NRC. 1996b. Review of Systemization of the Tooele Chemical Agent Disposal Facility. National Research Council. Committee on Review and Evaluation of the Army Chemical Stockpile Disposal Program. Washington, D.C.: National Academy Press.

NRC. 1996c. Public Involvement and the Army Chemical Stockpile Disposal Program. National Research Council. Committee on Review and Evaluation of the Army Chemical Stockpile Disposal Program. Washington, D.C.: National Academy Press.

OTA (Office of Technology Assessment). 1992. Disposal of Chemical Weapons: An Analysis of Alternatives to Incineration. Washington, D.C.: Government Printing Office.

Roberts, N.H., W.E. Vesely, D.F. Haasl, D.F. Goldberg. 1981. Fault Tree Handbook (NUREG-0492). January, 1981. Systems and Reliability Research, Office of Nuclear Regulatory Research. Washington, D.C.: U.S. Nuclear Regulatory Commission.

Smithson, Amy E. 1993. The Chemical Weapons Convention Handbook. Washington, D.C.: The Henry L. Stimson Center.

U.S. Army. 1987. Chemical Stockpile Disposal Program, Risk Analysis of the Disposal of Chemical Munitions at Regional or National Sites. SAPEO CDE-IS-87008. Aberdeen Proving Ground, Md.: U.S. Army Program Manager for Chemical Demilitarization.

U.S. Army. 1988. Chemical Stockpile Disposal Program Final Programmatic Environmental Impact Statement (FPEIS). Aberdeen Proving Ground, Md.: U.S. Army Program Manager for Chemical Demilitarization.

U.S. Army. 1994. Tooele Chemical Agent Disposal Facility Quantitative Risk Assessment Methodology Manual, December, 1994. Aberdeen Proving Ground, Md.: U.S. Army Program Manager for Chemical Demilitarization.

U.S. Army. 1995a. Anniston Chemical Agent Disposal Facility Phase I Quantitative Risk Assessment. SAIC-95/2542. Edgewood, Md.: U.S. Army Program Manager for Chemical Demilitarization.

U.S. Army. 1995b. Tooele Chemical Agent Disposal Facility Risk Management Plan, April 1995 (draft). Aberdeen Proving Ground, Md.: U.S. Army Program Manager for Chemical Demilitarization.

U.S. Army. 1996a. Pueblo Chemical Agent Disposal Facility Phase I Quantitative Risk Assessment. SAIC-96/2602. Aberdeen Proving Ground, Md.: U.S. Army Program Manager for Chemical Demilitarization.

U.S. Army. 1996b. Umatilla Chemical Agent Disposal Facility Phase I Quantitative Risk Assessment. SAIC-96/2601. Edgewood, Md.: U.S. Army Program Manager for Chemical Demilitarization.

U.S. Army. 1996c. Tooele Chemical Agent Disposal Facility Quantitative Risk Assessment. SAIC-96/2600. Aberdeen Proving Ground, Md.: U.S. Army Program Manager for Chemical Demilitarization.

U.S. Army. 1996d. Tooele Chemical Agent Disposal Facility Quantitative Risk Assessment. SAIC-96/2600 (Summary Report). Aberdeen Proving Ground, Md.: U.S. Army Program Manager for Chemical Demilitarization.

U.S. Army. 1996e. Chemical Agent Disposal Facility Risk Management Program Requirements. Aberdeen Proving Ground, Md.: U.S. Army Program Manager for Chemical Demilitarization.

U.S. Army. 1996f. Evaluation of the Carbon Filter System for Pollution Abatement Systems in Chemical Agent Disposal Facilities: Methodology for Evaluating Risks. December 1996. Aberdeen Proving Ground, Md.: U.S. Army Program Manager for Chemical Demilitarization.

U.S. Army. 1997a. Blue Grass Chemical Agent Disposal Facility Phase I Quantitative Risk Assessment. SAIC-96/1118. Aberdeen Proving Ground,

Md.: U.S. Army Program Manager for Chemical Demilitarization.

U.S. Army. 1997b. Pine Bluff Chemical Agent Disposal Facility Phase I Quantitative Risk Assessment. SAIC-96/1120. Aberdeen Proving Ground, Md.: U.S. Army Program Manager for Chemical Demilitarization.

U.S. Army. 1997c. A Guide to Risk Management Policy and Activities. 176-009. January 1997. Aberdeen Proving Ground, Md.: U.S. Army Program Manager for Chemical Demilitarization.

U.S. Army. 1997d. Pollution Abatement System Carbon Filter Simulation Model Development. January 1997. Aberdeen Proving Ground, Md.: U.S. Army Program Manager for Chemical Demilitarization.

U.S. Department of Labor. 1995. Census of Fatal Occupational Injuries. Bureau of Labor Statistics. Washington, D.C.: U.S. Government Printing Office.

U.S. NRC (U.S. Nuclear Regulatory Commission). 1975. Reactor Safety Study. WASH-1400, NUREG-75/014. Washington, D.C.: U.S. Nuclear Regulatory Commission.

U.S. NRC. 1990. Severe Accident Risks: An Assessment for Five U.S. Nuclear Power Plants. NUREG-1150. Washington, D.C.: U.S. Nuclear Regulatory Commission.

Utah DSHW (Division of Solid and Hazardous Waste). 1996. Tooele Chemical Demilitarization Facility Screening Risk Assessment. EPA I.D. No. UT5210090002. Salt Lake City, Utah: Department of Environmental Quality.

Whitacre, C.G., J.H. Griner, III, M.M. Myirski, and D.W. Sloop. 1987. Personal Computer Program for Chemical Hazard Prediction (D2PC). Technical Report CRDEC-TR-8702 1, January 1987. Aberdeen Proving Ground, Md.: U.S. Army Chemical Research, Development and Engineering Center.

APPENDICES

APPENDIX A

Perspectives on Risk, Risk Assessment, and Risk Management

This appendix is provided to develop a context for understanding risk assessment results. The initial sections deal with fundamental concepts of risk, risk assessment, and risk management. The latter sections explain in technical terms the meaning of the risk measures used for the DCD/TOCDF QRA.

TERMINOLOGY AND DEFINITIONS

This section provides definitions of terms used in the report, as well as terms used in the risk assessments. When appropriate, examples are provided.

Hazard is a possible source of danger.

Receptors are people, environmental components, or physical property exposed to a hazard.

Exposure is an opportunity for a hazard and a receptor to interact, creating an at-risk situation.

Risk is the possibility or probability that an undesirable outcome (e.g., damage, injury, or fatality) might result during, or as a consequence of, an activity or event that involves a hazard.

Risk assessment is a process focused on assembling and integrating relevant data to provide a quantitative (numerical) estimate of the probability of a particular outcome or range of outcomes.

Risk management is a decision making process focused on balancing alternative strategies and consequences associated with risk reduction and a process for implementing those decisions.

Voluntary risk is a risk that is known and understood (either quantitatively or qualitatively) by an individual(s) who has decided to accept that risk. Examples are sunbathing, driving an automobile, or smoking cigarettes.

Involuntary risk is a risk that may or may not be known or accepted by an individual but is imposed upon him. Examples are air pollutants emanating from a chemical manufacturing facility or from automobiles on a busy highway and radon seeping into basements from underlying bedrock.

RISK ASSESSMENT: AN ILLUSTRATIVE EXAMPLE

Risk

Consider a concrete sidewalk with a large, vertically displaced crack in it. The crack presents a **hazard** (source of danger) to **receptors** (people) who are **exposed** to it (i.e., people who use the sidewalk) because there is a possibility that they may trip over the crack. Thus, there is a **risk** to users of the cracked sidewalk. If such an event occurs (actual exposure), the sidewalk user may fall and be injured, or even killed (**risk consequences**). Users who are aware of the crack and choose to use the sidewalk are subjecting themselves to **voluntary risk**; users who are unaware of the crack and use the sidewalk (e.g., on a dark night) are subjected to **involuntary risk**. Voluntary users may choose to step over the crack (**risk management**). Ultimately, the owner of the sidewalk may choose to repair the crack and eliminate the risk altogether (again, **risk management**).

This simple example is **qualitative** in nature because it acknowledges the existence of a risk but does not consider the probability (or chance) of an actual event. In many situations, qualitative knowledge that a risk exists is sufficient for understanding and decision making (e.g., jaywalking on a busy street). Other situations, such as the destruction of chemical weapons, require in-depth understanding and quantitative analysis because of their complexity.

Extending the Example to Risk Assessment

A risk assessment is a process for developing quantitative (numerical) estimates of risk. In its simplest (generic) form, a risk assessment can be viewed as a four-step process. (Note that the NRC (NRC, 1983, 1994) and others have proposed alternative formulations.)

Hazard identification is the first step in the process, and as the term implies, it is concerned with documenting a hazard or hazards associated with a particular condition.

Consequence evaluation considers each hazard and the magnitude and likelihood of possible impacts on the receptor. A thorough analysis of failure or event sequences that can lead to the consequence allows estimates of both the likelihood and magnitude of the failure. For risk assessments involving toxic hazards, consequence evaluation is frequently referred to as dose-response evaluation, i.e., the evaluation of the human health effects from various doses of specific toxic materials.

Exposure assessment is an attempt to quantify the magnitude of possible (or actual) exposures, the pathways for exposure, the duration of exposures, and the size and nature of exposed population(s).

Risk determination combines the results of the consequence evaluation and exposure assessment to generate quantitative estimates of risk for each hazard and for all exposed populations.

The example of a cracked sidewalk illustrates the risk assessment process:

Hazard identification is a straightforward process that involves the simple visual observation of the fact that a tripping hazard exists because of the large, vertically displaced crack in the sidewalk. **Consequence evaluation** is somewhat more involved because it requires that consideration be given to all possible outcomes of a person actually tripping over the crack. A partial list of possible outcomes, in order of increasing severity, follows:

- person trips, loses balance, recovers
- person trips, bruises toe
- person trips, sprains ankle
- person trips, falls, breaks wrist
- person trips, falls, strikes head, dies

This is a simple set of possibilities. One could posit a more complex set of events, such as one person starts to trip, and a second person trips while trying to help the first. One of the two people falls into the path of a bicycle rider who falls off the bicycle. Fault trees are used to keep track of complex risk events.

Given a knowledge of the risk consequences, one can assign probabilities to the consequences of interest for purposes of risk assessment. In this example, death will be used as the consequence of interest. Death is often chosen as the consequence of interest in risk assessments because it can be clearly defined. If injuries are included, they might range from minor injuries to serious injuries that require longer periods of recovery. Sources of probabilities of occurrence can be either actual data (historical data from similar situations) or assumed values. In this example, the risk assessor would need to know: (1) the percentage of sidewalk users who actually trip over the crack; (2) how many of those who trip fall; (3) how many of those who fall hit their heads; and (4) how many of those who strike their heads die as a result. Sometimes no precisely relevant data are available. In those cases data may come from similar situations, theoretical models, or experts with relevant experience.

The goal of risk assessment is to produce realistic results that reflect the existing uncertainties. Often, "conservative" (upper limit) values are used to simplify problems. Conservative values set an upper bound on risk (i.e., the actual risk is expected to be less than these values); often an upper limit is sufficient for effective decision making. For the sake of simplicity, simple assumptions will be used in the example.

The most conservative answer to the question of the

percentage of sidewalk users who would trip over the crack would be 100 percent. But empirical knowledge tells us that this is unrealistic. A more realistic (but still conservative) number might be one in 50 sidewalk users (2 percent). Another assumption will be used to answer the second question, i.e., how many of those who trip will fall. Here, the assumption is that half of those who trip will fall (i.e., one in two). For the third question, how many of those who fall hit their heads, the assumption is one in 10; and for the fourth question, how many of those who strike their heads will die as a result, it is assumed that the chance of death from striking one's head on a sidewalk is one in 1,000.

Exposure assessment could establish that 10,000 different people might use the sidewalk at one time or another (based on nearby population data) and that the average is 1,000 people per day. (At this phase of the risk assessment, it might be necessary to further characterize the exposed population in terms of age, sex, weight, height, physical condition, etc., to make a truly comprehensive assessment because any one or any combination of these factors could influence the outcome.) The magnitude of the exposure is identical in this example for all exposed individuals because the crack is stationary and the size is constant.

To summarize, hazard identification, consequence evaluation, and exposure assessment have established the following:

- A hazard exists for users of the cracked sidewalk.
- Ten thousand people are potentially exposed to the hazard at some time (e.g., over a period of a year or two).
- One thousand people are exposed to the hazard each day.
- One person in 50 who use the sidewalk will actually trip.
- One in two individuals who trip will actually fall.
- One person in 10 of the individuals who fall will strike his or her head.
- One person of the 1,000 individuals who strike their heads will die as a result.

Risk determination, the final step in the risk assessment process, integrates the preceding information and develops a quantitative estimate of actual risk to sidewalk users. The probability (chance) that a user of the cracked sidewalk will trip, fall, and die can be developed by multiplying the probability of tripping (1 in 50) by the probability of falling (1 in 2) by the probability of striking one's head (1 in 10) and by the probability of dying from striking one's head (1 in 1,000). The result of this multiplication is 1×10^{-6}, or 1 in 1 million. In other words, mathematically the probabilities can be expressed:

$.02$ (1 in 50) $\times .5$ (1 in 2) $\times .1$ (1 in 10) $\times .001$ (1 in 1,000) $= .000001$ (1 in 1,000,000)

One interpretation of the result is that the probability is 1 in 1 million that an individual will die as a result of using the cracked sidewalk (**individual risk**). This probability is the same as the risk for a single use of the sidewalk by an individual. Another interpretation of the risk estimate is that 1 in 1 million people who use the sidewalk can be expected to die as a result (**societal risk**). Exposure assessment data (1,000 sidewalk users per day) can be used to calculate the probability of death per unit of time (day, week, month, year, etc.). This calculation yields an average of one death every 1,000 days (2.7 years) or 0.37 deaths per year or 0.001 deaths per day (other measures of societal risk). Several other probabilities or risk estimates could also be calculated.

This example involves a situation where an exposed hazard (the crack in the sidewalk) actually exists. In many situations, risk assessments must determine the probability that an internal or external event will create a new hazard or release a constrained hazard. These events are called **initiating events**. In the case of the cracked sidewalk, initiating events could be tree roots growing under the sidewalk, freezing and thawing during the winter causing the sidewalk to crack and buckle, or an earthquake.

To continue the example, assume that seismological data indicate that the frequency of earthquakes of sufficient magnitude to cause a sidewalk to buckle in the homeowner's geographical region is one in every 100 years. It may be further assumed that the chance of a homeowner's sidewalk buckling is 1 in 10 (as opposed to some other sidewalk) as the result of an earthquake.

When the risk estimate is considered in the light of initiating events that could produce the hazard, the overall probability becomes about 1,000 times less or 1 in 1 billion. Alternatively, one may think in terms of two related probabilities: (1) the probability of a hazardous condition being produced of 1 in 1,000; and (2) the probability of death from the hazard of 1 in 1 million, which is manifest only if the initiating event occurs.

Extending the Example to Risk Management

The selection of risk measures to be presented as the output of the risk assessment process depends on the objectives of the risk assessment. Ideally, the results of risk assessments are used by interested and potentially affected parties to make judgments and risk management decisions.

Continuing the example, suppose the sidewalk in question is a residential sidewalk and the owner of the residence is aware of the risk assessment results. The homeowner may decide that a risk of 1 in 1 million is acceptable and decide to do nothing. Conversely, he or she may decide to engage in a risk management program to reduce the risk to sidewalk users.

Among the risk management options would be posting signs to warn sidewalk users of the hazard, roping off the cracked area and requiring users to detour around it, building a ramp over the crack, or repairing the damaged area. Each of these options is accompanied by its own risks. For example, people who walk around the crack may trip over a tree root or step off a curb and sprain an ankle.

RISK ASSESSMENT FOR A COMPLEX FACILITY

This simple example helps to put the basic ideas of risk assessment in focus. However, it can be deceptive because, in the real world, especially in complex situations where quantitative risk assessments are used, neither the risks nor the useful models of risks nor the presentation of results is this simple. To move from the simple example to the models and results for the DCD/TOCDF operation requires more precise (mathematical) notions of probability, uncertainty, risk, and risk analysis. It also requires more complex models and carefully developed data, rather than simple assumptions, to describe the risk. This section includes expanded ideas for risk assessments and an overview of current risk assessment and risk management practices. Details of the risk modeling techniques used for DCD/TOCDF, along with a summary of the results from the DCD/TOCDF-specific risk assessments, can be found in Chapter 2. The following two sections, Probability and Uncertainty: A Modern Tower of Babel, and Hazard, Safeguards, and Risk, are for readers interested in the technical details that underpin the methods and results of risk analysis. They explain the meaning of the risk measures (see Chapter 2 of the main report) used for DCD/TOCDF in technical terms.

Probability and Uncertainty: A Modern Tower of Babel

The concepts of probability and uncertainty are relatively straightforward and are the basic vocabulary of risk assessment. However, the language describing these concepts has become confused and garbled, primarily because people use them to describe different concepts or use different words to describe the same concept. It is tempting to ignore all this and invent a new language, but that has been done several times in the past and has only contributed to the problem. To begin with a word, consider *probability*. Is probability (P) a measurable quantity from the real world or a way to calibrate an internal, mental, state of knowledge? Does it matter?

The notion of probability as a measurable parameter of the physical world (or at least as the subject of a conceptual experiment) is known as the relative frequency interpretation of probability. In this view, the probability of failure (P_F) has been represented as

$$P_r = \lim_{n \to \infty} \frac{F}{n} \qquad (1)$$

where F is the number of failures in n trials, i.e., the probability of an event is interpreted as the relative frequency of occurrence of that event in the long run. This concept has been known as probability, classical probability, frequentist probability, and frequency. It is now possible to construct mathematical formalisms to examine the behavior of *random variables* in a wide range of contexts (Cramér, 1946; Fisher, 1990; von Mises, 1957).

The second notion of probability requires acknowledging that, even if an objective, real world probability exists "out there," it may never be measured precisely. There is an element of uncertainty due to lack of knowledge. In this sense, probability becomes a structured scale (over the range of 0 to 1) that calibrates state of knowledge in a meaningful way. Various researchers have provided methods for constructing and calibrating this scale (de Finetti, 1974; Jaynes, 1996). With this scale, probability becomes a measure of what is in our heads rather than a measure of what is "out there." This concept has been known as probability, personal probability, subjective probability, prevision, degree of belief, Bayesian probability, and state of knowledge (de Finetti, 1974; Jaynes, 1983; Jeffreys, 1961; Lindley, 1965; Savage, 1954). This view of probability has led to the development of methods for treating decision making under conditions of uncertainty and for addressing a wide range of open-ended physical problems where substantial uncertainty exists, i.e., most physical problems risk assessment attempts to address.[1]

Finally, does all this matter? Sometimes, when substantial facility-specific data are available, for example, either way of thinking leads to the same numerical results. Sometimes the difference matters very much, both numerically and philosophically. The battle over the correct interpretation of probability has been raging for more than 100 years (Krüger, Daston, and Heidelberger, 1990). At times, it has been a bitter conflict. Even though the two kinds of uncertainty are not difficult to tell apart, the protagonists have managed to ignore each other's ideas, often demonstrating only that "If you accept my definition of probability, then my opponent's calculations of probability are flawed." Two great opponents in this debate, R.A. Fisher and H. Jeffreys, vigorously debated the philosophies underlying their theories and then calculated the same results because each was wise enough to recognize the requirements of the specific problem at hand and adapt his methods to accommodate the proper question (Lane, 1980).

In a recent attempt to reconcile these conflicts, several prominent workers in quantitative risk assessment (also called probabilistic risk assessment and probabilistic safety assessment) have suggested a return to unambiguous language. They call the uncertainty associated with the random nature of the events being modeled *aleatory uncertainty* and the uncertainty associated with the analyst's state of knowledge about the processes that govern that randomness *epistemic uncertainty*. The aleatory uncertainty captures variability that is observed but is beyond the explanation of the physical models used in the analysis. The epistemic uncertainty allows for our lack of knowledge (i.e., lack of observation).

If these two types of uncertainty are combined improperly, the result can be an underestimation of epistemic uncertainty (Mosleh et al., 1994). One example from the DCD/TOCDF QRA pointed out by the Expert Panel "is the variability in inventory that results from the operational practice of loading the container handling building for nighttime operation, which was handled as an uncertainty in the inventory. This, however, is a random factor with respect to the initiating event occurrence, and would be better reflected as an aleatory distribution on the source term" (MITRETEK Systems, 1996). The DCD/TOCDF QRA has adopted *aleatory/epistemic* language for uncertainty and uses the word probability in all cases (frequency and state-of-knowledge concepts).

Form of the Results

The results of analyses that support risk assessment include, in the language of the DCD/TOCDF QRA, probabilities of certain events (e.g., the frequency of initiating events, such as dropping munitions, are expressed as probabilities over the relevant agent destruction campaigns). Figure A-1 illustrates two types of presentation. Type 1 is a point estimate, i.e., a single number that characterizes the result. Here the point estimate of the probability is the mean value ("the probability" to many practitioners). The mean value is the weighted sum of all possible values (the integral is for continuous distributions) and is considered the most appropriate point estimate for summary purposes. Other point estimates include the median (the 50th percentile, for which half of the possible values lie below the median and half above) and the 95th percentile (95 percent of the possible values lie below and 5 percent above).

In Figure A-1, the Type 2 presentation is the full expression of uncertainty. The Type 2 curve is known as a density function, where the probability of lying in any interval is the integral over that interval. The point estimates discussed above are summary "statistics" calculated from the density function.

---

[1] The reader interested in gaining additional experience with probability calculations is referred to Feller (1968). For details on carrying out Bayesian calculations, see Kaplan and Garrick (1979). For a less technical but intriguing history of the ideas of probability and risk, see Bernstein (1996).

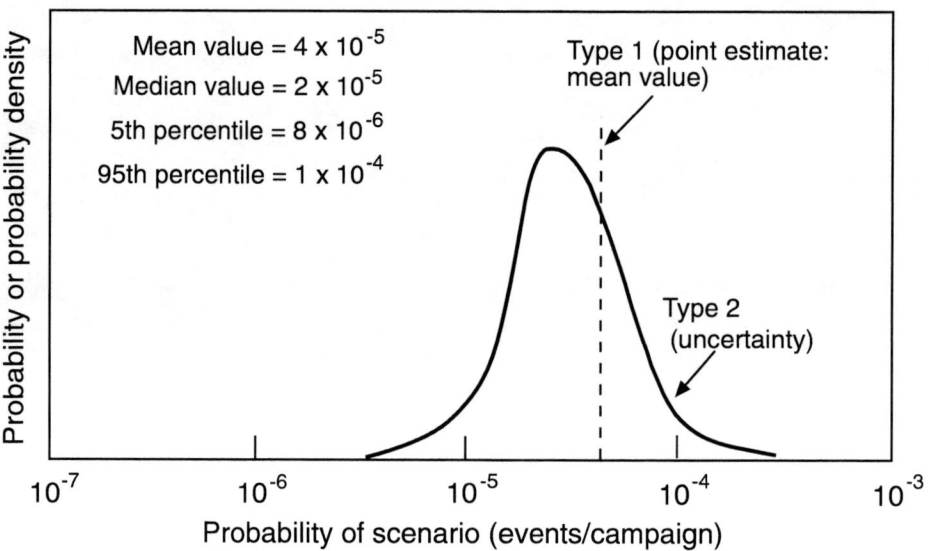

FIGURE A-1 Form of the results: scenario probability.

Figures A-2a and A-2b illustrate the differences in representation of aleatory and epistemic uncertainties. Figure A-2a is a representation of pure epistemic uncertainty. Here some event will either surely happen (p = 1) or fail to happen (p = 0) under certain conditions. Because it cannot happen only some of the time, only two values are possible: yes (p = 1) or no (p = 0). Therefore, the density function has two spikes, one at p = 0 and one at p = 1. Although there is only one correct answer, there is state-of-knowledge uncertainty about which one is correct. As a specific example, *identical* stacks of munitions will either fall or remain standing following *identical* shaking in response to a *particular* earthquake. Suppose that current knowledge from observations and calculations (the particular earthquake and shaking have not yet occurred) is that the state-of-knowledge probability is 0.3 that a particular stack will not fall (p = 0) and a complementary probability of 0.7 that it will (p = 1). (In terms of a density curve, this amounts to a Dirac delta function of value 0.3 at 0 on the scale of probability of the event "stack does not fall.") This is pure epistemic uncertainty. Once the

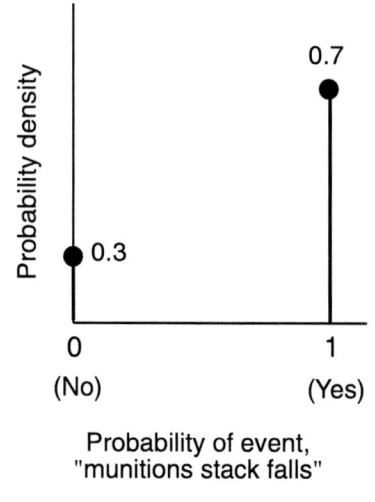

a. Pure Epistemic Uncertainty

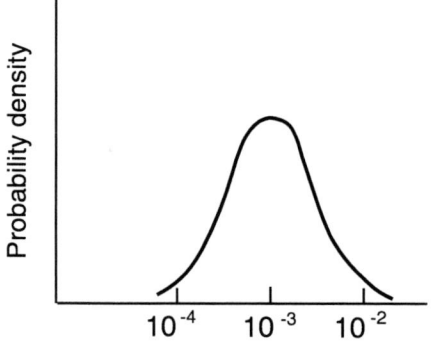

b. Pure Aleatory Uncertainty

FIGURE A-2 Aleatory and epistemic uncertainty.

particular earthquake and shaking occur, it will be known with certainty either that the stack falls or that it remains standing. No uncertainty remains.

Figure A-2b is an example of pure aleatory uncertainty. Here an event will happen only some of the time under particular conditions. The resulting probability density curve represents the fraction of the time (relative frequency) that the event occurs (e.g., a particular machine fails during one hour of operation). In the figure, the most likely value is 1×10^{-3}/hr. However, randomness among similar machines means that some fail at a rate of 1×10^{-4}/hr and some as often as 1×10^{-2}/hr.

Finally, pure aleatory and epistemic cases are rare (or may never occur), which has given rise to the strident arguments from those whose work is aimed at solving one problem or the other. Consider the example of stacks of munitions subjected to an earthquake. There are many reasons the uncertainty of this problem does not disappear following the "experiment" of encountering a particular earthquake and shaking. There is an element of randomness in the construction and positions of the stacks that will affect their response to the shaking. Some parameters of the earthquake that affect shaking cannot be modeled and tracked within a quantitative risk assessment (e.g., vertical and horizontal displacement, frequency, and time history), and other intractable factors link shaking to the earthquake itself. Alternatively, consider the aleatory failure rate for the machine. Machines fail in particular ways after particular shocks and stresses. Increased knowledge and modeling of these factors could reduce the randomness. At the same time, even if absolutely identical machines could be constructed, they would be installed in different facilities by different workers, operated under different conditions, and maintained by workers following local policies. Thus, state of knowledge is embedded in most, if not all, cases that at first appear to be purely aleatory.

Hazard, Safeguards, and Risk

A **hazard** was defined earlier as a possible source of danger. A material may be a hazard because it is toxic to receptors (e.g., a chemical agent), because its potential energy (chemical, mechanical, electrical, or nuclear) could be released causing direct or indirect harm to receptors (e.g., an explosive reaction or an earthquake that topples a stack of munitions), or because its presence could cause the receptors' own activity to lead to harm (e.g., crack in the sidewalk that causes a passerby to trip and fall). The attributes of a hazard must be characterized and may be possible to control, such as mass, toxicity, energy content, shape, and size. However, the presence of a hazard does not guarantee harm.

Safeguards[2] stand between hazards and receptors. The term safeguards is used here to describe any physical or procedural barrier (designed or natural) that protects receptors from a hazard. Chemical agents may be stored in sturdy steel ton containers, making exposure to workers or the public quite unlikely. A large propane tank, potentially subject to destructive earthquakes, can be maintained with a limited volume of propane to minimize the tank's structural response to the earthquake, thereby reducing the chance of rupture and subsequent explosion. A chemical processing facility can be located far from large population centers. Signs, lights, and physical barriers can warn walkers of the presence of a crack in the sidewalk. Thus, safeguards can be introduced to control risk. The risk (chance of an undesired outcome) is a function of both the hazard and the safeguards.

Simple risks can be analyzed qualitatively. In the sidewalk example, even the simple analysis may be perceived as tutorial overkill. No one needs to calculate the probability of death from tripping to know that they should protect their neighbors from this hazard and repair the crack. But this example can quickly become complicated. An organization that owns miles of sidewalks may not even know a crack has developed, or, although they know that cracks must exist, they may not know if cracks severe enough to pose a hazard exist or where they are. In this situation, quantitative notions of risk and hazard can provide managers with useful information for controlling the risk to neighbors, employees, and residents. They will need to characterize cracks that pose a hazard (e.g., by length, breadth, and vertical displacement). They will need to examine the range of initiating events that can cause cracks (e.g., growth of tree roots, thawing-freezing cycles, trucks crossing sidewalks) and the probability of each event.

---

[2]The term **safeguards** may have special meaning in some industries. Although this may cause some confusion, the discrepancy is not charged with the same historical baggage as the discrepancy about **probability**.

They will need to define the risk in terms of probabilities and consequences. In other words, they will need to perform a risk analysis and present the results in quantitative form. If the organization communicates its concern to neighbors and sets up a simple system for reporting cracks to a central location, a good source of information will be available.

The situation is substantially more complicated for a large chemical processing facility, such as the TOCDF, where there are numerous hazards in many locations and where a variety of processing activities may provide hazards with multiple pathways to workers and the public. Simultaneously, these hazards can be affected by a wide range of safeguards.

The first requirement in analyzing complex situations is to define risk in a way that will be meaningful to managers, individuals (workers and the public), and emergency planners. Historically, the first risk measure proposed for risk studies was the "expected consequences." This widely used measure is the basis (limit or requirement) specified in many environmental regulations and is the sole product of most health risk assessments. Expected consequences are a mathematical construct rather than a characteristic of specific accidents. The expected risk is defined as

$$expected\ risk = \int_0^\infty x \cdot f(x) dx \qquad (2)$$

where x is the consequences, $f(x)$ is the probability density function, and $f(x)dx$ is the probability that x lies in the small region between x and $x+dx$. (Note that in the case of discrete consequences, such as fatalities, the mathematics become discrete rather than continuous, i.e., 2.5 deaths is meaningless). Because of this formulation, it is sometimes said that

*risk = probability **times** consequences*

To understand this measure, consider a clone of DCD/TOCDF (i.e., consider a hypothetical, very large set [millions or billions] of identical DCD/TOCDF sites [including the nearby population and surroundings]). If the millions of identical sites were operated for their actual lifetimes (estimated as 7.1 years) and if all the accidents causing deaths at all the sites in the clone were tabulated, then the average number of deaths over the millions of sites would be the "expected fatalities" from operating DCD/TOCDF. This average is typically very small, perhaps one-third of a death or 1/10,000 of a death, or even less, because in a well-designed facility, accidents involving fatalities are extremely unlikely. The expected risk is an average over all possibilities rather than a result that is "expected" in the ordinary sense. Moreover, there is only one DCD/TOCDF and, if it has an accident at all, that accident will have one outcome, and that outcome will not be the "expected" number of fatalities. It will be one specific outcome from the range of possible outcomes (e.g., no deaths, 1 death, 10 deaths, or perhaps 100 deaths).

Although the risk measure, *expected consequences*, is often used, it may not be an adequate measure of risk. It does provide a rough summary of the level of risk posed by a facility. However, because it is a high-level average, many important details are obscured. Note that the following three facilities would have the same "risk," in terms of the expected number of deaths, i.e., 0.0045 or 1/220:

- a facility with risk "dominated" by one accident that would kill 300,000 people with a probability of 1.5×10^{-8} of that accident occurring over the lifetime of operation (i.e., it almost certainly will not occur, but, if it does, it will overload local medical facilities, destroy nearby communities, damage the economic base of an entire state, and be an internationally recognized disaster)
- a second facility with risk dominated by two accidents: one that would kill one person with a probability of 5×10^{-4} (very unlikely, but such events have happened); a second one with only a 1 percent chance of killing one person and a probability of 0.4 (this accident is about as likely as tossing heads with a coin, but the consequences are as unlikely as death during a medical operation with general anesthesia; overall, accidents of this severity are commonplace)
- a third facility whose risk is dominated by one accident that would kill 10 people with a probability of 4.5×10^{-4}

These three risks have the same number of expected fatalities, but they are in fact very different risks, both in terms of the likelihood that they will occur and in terms of the magnitude of the impact if they occur. Presenting the risk in terms of

*risk = probability **and** consequences*

rather than the *probability **times** consequences* summary of expected consequences, provides a more thorough understanding and improves the chances of effective risk management.

A display format for summarizing the presentation of *probability **and** consequences* is known as the risk curve or *risk profile*. For fatalities, the risk profile displays the probability of an accident involving "*x* or more" fatalities as a function of *x*, the number of fatalities. The three risk profiles for the three simple results above are all shown on Figure A-3.

Moving from the simple examples with one or two dominant accidents to a risk assessment of a complex site/facility like DCD/TOCDF means the risk curve will be more complex. To address the more complex situation, it helps to lay out the risk in a general format. See Kaplan et al. (1981) for further explanation of the notation used below. In this format, risk is simply the answer to three questions:

1. What are the scenarios that can cause damage? Call each one S_i.
2. What is the frequency of the scenario? Call it Φ_i.
3. What are the consequences? Call them X_i.

The answers to those questions come in the following form:

"An" Answer $\quad <S_i, \Phi_i, X_i>$
Set of Answers $\quad \{<S_i, \Phi_i, X_i>\}$
Complete Set $\quad \{<S_i, \Phi_i, X_i>\}_C$

and the risk is

$$R = \{<S_i, \Phi_i, X_i>\}_C \qquad (3)$$

which includes $\quad S_0 =$ "As Planned Scenario"

So a risk assessment is a list of all triplets $<S_i, \Phi_i, X_i>$. The art of risk assessment is in structuring the scenarios in a way that facilitates analysis and computation. The tools for this process include logic modeling (discussed in Chapter 2) and mechanistic calculations based on science and engineering. Uncertainty enters this picture in terms of completeness (have all the important scenarios been identified), in terms of frequency (events per year), and in terms of consequences. Completeness can be directly addressed in limited scope risk assessments in several ways, including making allowances for scenarios that are knowingly omitted (Bley, Kaplan, and Johnson, 1992).

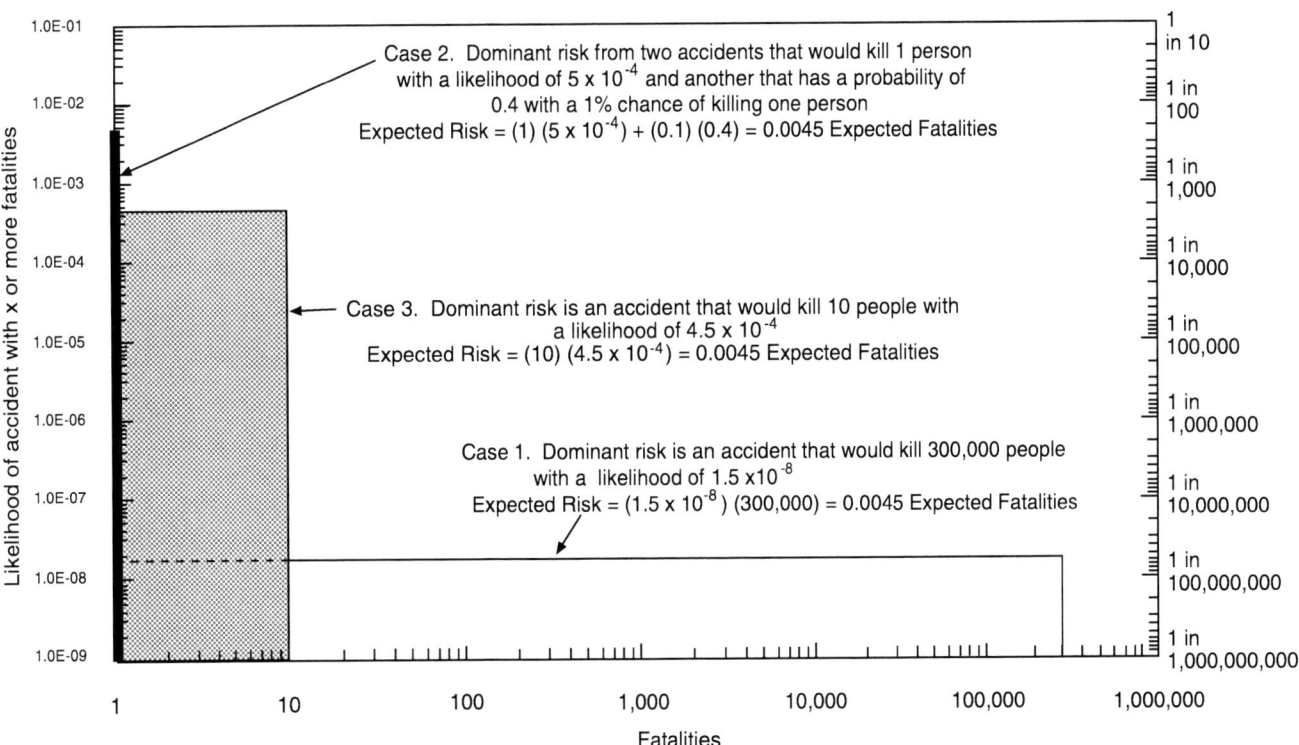

FIGURE A-3 Risk profiles with the same expected risk.

For full scope risk assessments, every effort is made to be complete by structuring a search for initiating events, by reviewing histories at similar facilities, by examining accident calculations, and by extensive reviews. Uncertainty in the consequences is generally considered by breaking up the range of possible consequences into a number of discrete possibilities, then subdividing the scenarios into many subscenarios, each with its own consequences. Following this process, all the uncertainty is contained in the estimate of the frequency or probability of the scenarios. At this point, consistent with the language of the DCD/TOCDF QRA, replace the frequency in equation (3) with the probability over the appropriate campaigns (p_i).

A large-scale risk assessment like the DCD/TOCDF QRA develops a very large number of scenarios. An easy way to understand the presentation of results, called a risk curve or a risk profile, is to think of the list of scenarios above, $R = \{<S_i, p_i, X_i>\}_C$, as a table in which the scenarios are rearranged in the order of increasing consequences:

$$X_1 \leq X_2 \leq X_3 \leq ... X_N$$

Add a fourth column showing the cumulative probability (P_i), i.e., uppercase P, as shown in Table A-1. Note that

$$P_1 \geq P_2 \geq P_3 \geq ... \geq P_N$$

so that P_i can be considered the probability of exceedance (the probability that the consequences are equal to or greater than the associated X_i).

When the points $<X_i, P_i>$ are plotted in Figure A-4, the result is a staircase function. Next note that the scenarios in Table A-1 are really categories of scenarios.

TABLE A-1 Scenario List with Cumulative Probability

Scenario	Probability	Consequences	Cumulative Probability
S_1	p_1	X_1	$P_1 = P_2 + p_1$
S_2	p_2	X_2	$P_2 = P_3 + p_2$
.	.	.	.
.	.	.	.
.	.	.	.
S_i	p_i	X_i	$P_i = P_{i+1} + p_i$
.	.	.	.
.	.	.	.
.	.	.	.
S_{N-1}	p_{N-1}	X_{N-1}	$P_{N-1} = P_N + p_{N-1}$
S_N	p_N	X_N	$P_N = p_N$

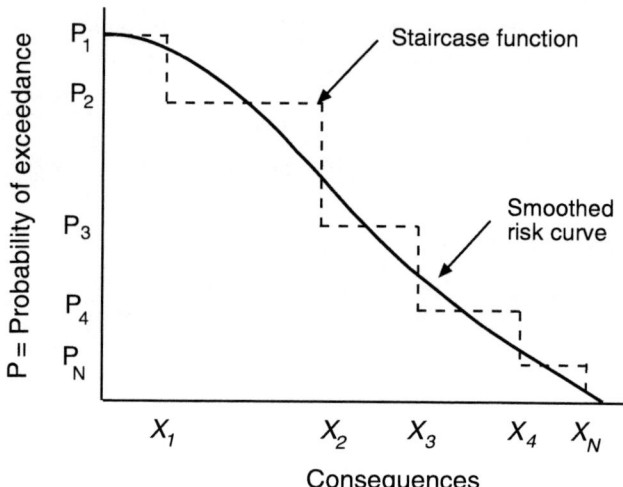

FIGURE A-4 Risk curve.

For example, the "munitions drop" event actually includes a large number of slightly different scenarios, each resulting in slightly different consequences. Thus, it could be argued that the staircase function should be regarded as a discrete approximation to a nearly continuous reality. If a smooth curve is drawn though the staircase, that curve can be regarded as representing the actual risk, and it is called the risk curve or risk profile.

Thus the meaning of the risk profile is clear. Turning to Figure A-5, the Type 1 (point value) risk curve is familiar. Here P_1 is the probability that the consequences are equal to or greater than the consequence X_1. The Type 2 risk profile addresses uncertainty. Here there is a family of risk curves (or a risk surface). Now the authors are p_3 confident (perhaps 95 percent) that consequences X_1 or greater are no more likely than $P_{1,3}$. They are p_2 confident (perhaps 50 percent) that consequences X_1 or greater are no more likely than $P_{1,2}$. Finally, they are only p_1 confident (say 5 percent) that consequences X_1 or greater are no more likely than $P_{1,1}$ (i.e., there is a 95 percent chance that they happen more often).

In most QRA studies at least two classes of consequences are considered—acute and latent health effects. Acute health effects involve immediate injuries and deaths. Immediate injuries associated with agent release at the TOCDF tend to be minor, reversible effects of very low-level exposures to nerve agent (e.g., watery eyes and runny noses). In comparison to deaths and latent cancer effects, immediate injuries are minor

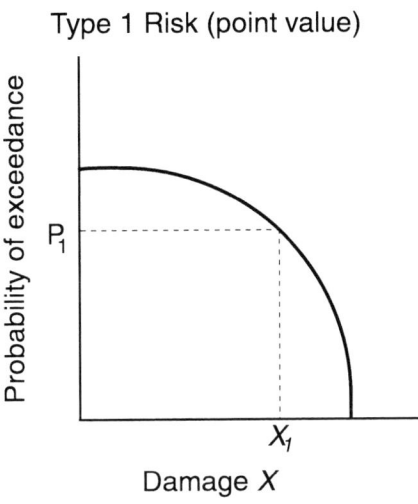
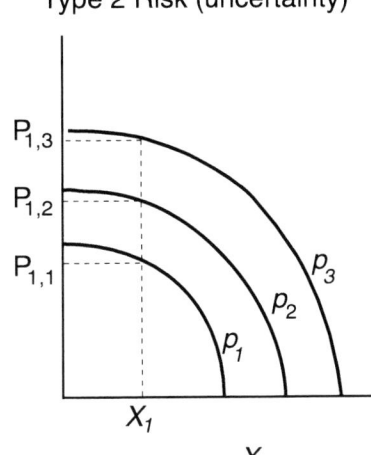

FIGURE A-5 Form of the results: risk profiles.

and are not reported in the DCD/TOCDF QRA. The most severe latent health effects are possible cancers from exposure to mustard. These cancers, if not properly treated, can become deaths many years later. Risk profiles shown in the main report are associated with immediate (acute) fatalities.

Some General Classes of Risk Analysis

The risk assessment, as developed above, is in the general format used in a QRA. It attempts to be complete, in the sense that the QRA attempts to quantify all scenarios that substantially contribute to the risk measures of interest to those who have chartered the study and to address all uncertainties. Not all QRAs are full-scope studies. In some QRAs, the analysis addresses only internal events or only external events. In many QRAs, the analysis addresses only accidents. Historically, this has been because the risk of death and serious injury *to the public*, which result only from accidents, has been the focus of the QRAs. The DCD/TOCDF QRA addresses both internal and external events, but only accidents involving agent. For the TOCDF, the risks from normal (non-agent) emissions and minor upset conditions are addressed in the health risk assessment (HRA).

HRAs generally involve a simplification of the basic model described above. Typically, they only examine risks from normal operation and mild upset conditions. The scope of HRAs has been prescribed by the Environmental Protection Agency (EPA), and, therefore, an HRA is rule-driven, rather than science-driven. Although there has been some criticism of this approach (NRC, 1994), it does have some advantages. Criteria are established defining what must be analyzed, how it must be analyzed, and the standards that must be met. This approach allows for a simpler analysis than a full risk assessment. Uncertainties are replaced by using conservative upper limit assumptions on releases of hazardous materials. Even with these conservative requirements, facilities can be engineered to meet EPA limits for releases of cancer-producing chemicals, and there seems to be wide acceptance of the HRA approach. Difficult questions of substantial uncertainty, such as the body's response to very low doses and the possibility and consequences of rare accidents, are not addressed. Questions of values and policy are embedded in the requirements and are, therefore, not revisited for every new application.

QRAs and HRAs are similar in many ways. Both could be called *facility-centered* risk assessments in that they focus on a single facility and are performed to manage (or regulate) that facility. Both evaluate the impact of facility operations on nearby populations and property. (Sometimes the QRA, like the DCD/TOCDF QRA, also evaluates the impact on workers.) Both are used to manage risk by changing facility design or operation and by managing emergency response practices. The primary differences are the types of risks that are examined and the treatment of uncertainty. The QRA examines accidents (and normal releases, if they contribute substantially to risk); the HRA examines normal releases and mild upset conditions. For DCD/TOCDF, the QRA focuses mainly on agent releases, while the HRA is concerned with emissions resulting from the destruction of agent and munitions. The QRA attempts to calculate a realistic result, including uncertainty, which permits management to consider the best estimate of the effects of changes on risk. The HRA calculates an upper limit on releases and health effects, forcing management to meet a pass/fail criteria.

Perception and Risk Assessment

In the past several decades, risk assessment and risk management have become major factors in making decisions involving potentially adverse consequences to society. During this period, risk-related concerns have also permeated the public consciousness. Because meaningful measures of risk can now be generated, people and organizations are being asked to take increased responsibility for the risks they impose upon themselves and others.

More awareness of the risk of an activity does not necessarily translate into an understanding that can be quantified. People may place different values on particular risks, depending on their personal views. Often, the perception of risk by significant segments of the general population has not progressed beyond the level of intuitive feelings based on personal experiences, culture, and mass media coverage (Piller, 1991). Whether risks have been quantified in terms of consequences and frequency of potential occurrence (e.g., one chance in 6,000 per year of being killed in an automobile accident in the United States) or are only vaguely perceived as detrimental influences to an individual, a family, or a community, the political and societal implications need to be examined in an orderly manner.

People have different ideas about which risks are acceptable. Some people may smoke but be afraid of skiing, or vice versa. These are voluntary risks that allow people to choose based on their personal perceptions of risks and benefits. Risks of certain diseases and natural disasters are largely involuntary, although people may take some preventive measures. Involuntary risks associated with a wide range of industrial activities are managed by society through codes, standards, regulations, economic considerations, and responsible behavior. For hazardous chemicals, such as pesticides or highly flammable or toxic materials, a high level of risk analysis is often desirable. For example, in the DCD/TOCDF QRA, each phase of activity is analyzed to determine how accidents might be initiated and progress.

Risk communication is a separate discipline. Risk analyses are very large integrated studies that can be difficult to understand. They involve many different kinds of expertise, modeling, and calculation. Expert input, often in the form of assumptions, is required to limit the scope of the modeling and to permit the models to include information on the boundaries of scientific knowledge. Communicating the content and results of risk assessments in ways that can be understood, that clarify the uncertainties, and that draw a fine distinction between facts and policies has proven to be difficult. Since the publication of the *Reactor Safety Study* (U.S. NRC, 1975) (the first full nuclear plant QRA, which was widely criticized for the presentation of results in the summary report), extensive research has been done on communicating risk results (NRC, 1989).

Effective communications have been hampered because three traditions (QRA, HRA, and risk communication) are involved, each with its own history, practitioners, and literature. Although some attempts have been made to reconcile them, including the formation of a technical society, the Society for Risk Analysis, they have remained largely separate. The same can be said of the world of practice. The three traditions have progressed rather independently of each other but have converged in the Army's Chemical Stockpile Disposal Program.

Earlier recommendations of the Stockpile Committee have urged that the risk assessments be integrated and combined with an effective risk communication and public involvement program to ensure that interested parties, such as the public, local and national government entities, and the Army, all understand the risks involved in continued storage and alternative methods of destruction. The present report reviews and comments on the QRA and HRA studies performed for DCD/TOCDF and on the tools established for managing the risk. This report provides perspectives on how the studies can be viewed and used in an integrated way. The committee hopes this report will help interested parties understand how to interpret and use the results of the risk analyses of DCD/TOCDF and other sites/facilities.

Risk Management

For a chemical agent and munitions disposal facility and its associated storage site, risk assessments of accidents (e.g., a dropped rocket), transients or upsets (e.g., stack agent release), and normal operations can be developed at different levels of detail depending on the available information and the intended use of the results. For example, the TOCDF HRA is a screening risk assessment based on conservative assumptions (overestimates) about emissions and is intended to demonstrate that the risk is below permit requirements. The TOCDF QRA is a detailed, site-specific risk assessment (best estimate and full statement of uncertainty) intended to evaluate and facilitate management of the risks associated with accidents involving agent.

Risk assessments are intended to provide a quantifiable scientific basis for managing facility design and operations. Once the whole spectrum of risks has been quantified, it is possible to evaluate issues, such as whether or not maintenance of a spare piece of equipment has a significant impact on operational safety. A risk management plan that lays out the process for using risk assessment information within the overall plant management structure is essential to taking full advantage of a thorough risk assessment.

Risk management addresses such matters as proper interactions between managers responsible for controlling risks and the individuals on site and off site who are responsible for emergency preparedness and accident mitigation. A risk assessment identifies the major causes of risk and can be useful for developing options to reduce risk. For example, the risk assessment may be used for ordering the sequence of destroying particular weapons to reduce the stockpile risk as quickly as possible. Other areas where a risk management plan uses the results of a risk assessment in decision making include the management of change, performance evaluation, and incident investigation. Conversely, the information that derives from risk management can be used to refine and enhance the accuracy of a risk assessment. A more thorough discussion of the risk management process is given in the next section.

RISK MANAGEMENT PROCESS

Risk management can be described as the process by which risks are understood and controlled. All affected parties have roles to play in the risk management process at DCD/TOCDF. The Army is responsible for managing the chemical stockpile and its destruction. However, the Army's contractors, individual workers, local governments, and the affected public must all participate for the process to proceed efficiently and safely (NRC, 1996a). Risk management usually involves the following steps:

- understanding the risk (including identifying major contributors to risk)
- suggesting alternative ways to reduce risk
- evaluating risk reduction alternatives
- selecting preferred alternatives (including implementing decisions)

Step 1: Understanding the Risk

Understanding the results of risk assessment implies more than knowing the summary numerical results of the QRA and HRA. Understanding also requires knowing the details, including the assumptions, simplifications, and omissions, of the analyses. The results must be viewed in the full context of the risk assessment, as well as in the context of the actual safety performance of the plant. This must be accompanied by a thorough understanding of explicit and implicit uncertainties.

Understanding the results of the risk assessment also means knowing the significant contributors to risk, i.e., knowing how improved performance can reduce risk and how degraded performance can increase risk. The possible benefits are listed below:

- Managers and workers can develop options for reducing risk or for ensuring that risk does not increase. They can also consider how proposals for change affect risk.
- Workers, emergency response personnel, and others can better understand their personal risks and how best to protect themselves and each other.
- Emergency preparedness managers can focus their planning and training programs on the most important scenarios or sources of risk to the surrounding communities.

- State and local officials can provide more informed oversight in their decision making.
- Everyone can participate knowledgeably in the risk management process.

For example, risk from seismic events was found to be the dominant contributor to the risk of fatalities at DCD/TOCDF. The Army has modified operating practice to reduce one of the major seismic contributors (see Example 1 in Chapter 3). Emergency preparedness officials of the Chemical Stockpile Emergency Preparedness Program should also be aware of the nature of seismic risks and keep them in mind when developing and implementing their response plans.

The DCD/TOCDF QRA and HRA reports provide sufficient detail for understanding the risks associated with the Army's Chemical Stockpile Disposal Program (U.S. Army, 1996; Utah DSHW, 1996). See Chapter 2 for a summary of this information, which has also been presented in public meetings near the Tooele site.

Step 2: Suggesting Alternative Ways to Reduce Risk

Risk can be reduced through effective changes of equipment, activities of plant personnel, and emergency response capabilities. Uncertainties in calculated risks can be reduced by better understanding the factors affecting risk. Some examples of risk reduction alternatives follow.

Changes to Plant Hardware

These are obvious responses to risks involving plant equipment. These changes are often costly, however, and may involve retraining workers; therefore, other alternatives should also be considered, which may turn out to be more effective. Changes to plant hardware have been considered at the TOCDF and several have been implemented (see Examples 1 and 2 in Chapter 3).

Changes to Plant Procedures

Operating, maintenance, and emergency procedures, as well as related off-site emergency response procedures, can be effectively modified and improved to reduce risk. Care must be taken to ensure that neither the training of personnel nor the level of performance is adversely affected by frequent or poorly analyzed procedural changes.

Changes to Emergency Response Capabilities

Plans, preparations, and mitigation activities by the Chemical Stockpile Emergency Preparedness Program and other emergency response organizations can be revised or restructured to deal more effectively with the major identified contributors to risk. The relative risks associated with alternative responses can also be assessed.

Changes in Management Philosophy and Incentives

These can involve a wide range of activities. For example, changes in training that increase the knowledge and improve the skills or behavior of on-site and off-site personnel can improve performance. Another example would be changes in the criteria for performance evaluation and compensation that could alert both managers and workers to the relative importance of certain factors, such as safety, environmental performance, and productivity.

Changes in philosophy include management response to errors or other failures. If management response is punitive, then mistakes will be covered up. If the management goal is a high level of safety and environmental performance, and if sharing problems and near misses is seen as an opportunity for learning and improvement, then safer operations are more likely to result (Chess, Greenberg, and Tumuz, 1995; Ochsner, Chess, and Greenberg, 1995).

Changes in Organizational Culture

Management can also be proactive in establishing a culture throughout the organization that strives for the continuous improvement of safety and environmental performance in all aspects of operation.

Improvements in Knowledge to Reduce Calculated Risks

Reducing uncertainties often has a tendency to reduce calculated average risks because the average is strongly affected by possibilities associated with upper uncertainty bounds. Efforts to reduce calculated risks typically involve improvements in basic scientific knowledge and improvements in risk modeling.

Improved Basic Knowledge. Options for gathering or developing new information include extending the review of the scientific literature, eliciting opinions from experts, making more accurate mechanistic calculations, performing experiments and tests to determine new scientific information, and focusing analyses of performance data to find new insights into the behavior and interactions of plant conditions and workers.

Improvements in Risk Modeling. Risk models necessarily involve simplifications, approximations, and assumptions. Improvements in risk modeling are usually possible if analysts can refine their models by replacing worst-case assumptions with detailed analyses. In the initial phase of a risk assessment, it is often necessary to use conservative models that overestimate risks, expending the effort to be more accurate only on those parts of the analysis that have a significant impact on results. Thus initial estimates may exaggerate some risks. In some cases, additional risks are discovered through detailed analyses, especially if the range of possible uncertainties was not carefully considered. Improved data are also possible once a facility begins operations (e.g., the Johnston Atoll Chemical Agent Disposal System or the TOCDF) because models can be refined using facility-specific data rather than data from similar facilities.

Step 3: Evaluating Risk Reduction Alternatives

For every proposed change (in design, equipment, or procedures), it is necessary to assess the impact of the change on safety, ease of operation, environmental performance, public and worker health, short-term and long-term economic costs and benefits, schedules, regulatory compliance, political and public acceptability, and flexibility to respond to future mandated or voluntary changes (OTA, 1995). Changes that may have a significant impact on safety, health, or the environment need to be carefully assessed so that trade-offs and changes in risk are well understood. Changes can be suggested by managers, operators, inspectors, and other interested parties, and are often required by regulators. Evaluations are influenced by costs, schedules, and advances in technology.

Step 4: Selecting Preferred Alternatives

When considering alternatives for risk reduction, it is appropriate to consider making no change as an option, or at least as a yardstick, for comparison. The decision process involves matters of fact (e.g., changes in risk as calculated in the risk assessments); limitations on the facts (e.g., assumptions, approximations, and uncertainties); and matters of policy (e.g., how safe is safe enough, who should pay, and the value of trade-offs). There is no easy formula for weighing these factors, especially when trade-offs are involved. However, failure to give fair consideration to all of these factors can be a recipe for controversy and failure. Public participation is especially important when scientific facts and policy issues must be balanced. Who decides and through what processes decisions will be made involve extremely complex questions. In a given situation, the dynamic interaction of factors, such as legislative mandates, organizational philosophy, and public awareness and organizational involvement, dictate the basic framework of the answers to these questions.

In the past 15 years, formal tools for managing risk at technological facilities have substantially improved. Risk assessments and risk management systems are described in the literature for a variety of facilities in the electric utility, aerospace, transportation, and chemical process industries (NRC, 1996b). Technical conferences often devote numerous sessions to risk management and risk-based or risk-informed regulation. Risk-based regulations are founded on risk assessments; risk-informed regulations consider risk information along with other factors. Conference proceedings include many examples of risk management processes for a variety of industries (e.g., Vesely et al., 1995). Some risk limits are regulation driven. RCRA Part B regulations, for example, set limits on normal process releases from combustors. The U.S. Nuclear Regulatory Commission is about to issue new safety evaluation reports and standard review plans that formally implement a risk-informed regulation process for nuclear power plants as part of its previously announced policy on using risk analysis in regulation (U.S. NRC, 1995).

REFERENCES

Bernstein, Peter L. 1996. Against the Gods: The Remarkable Story of Risk. New York: John Wiley & Sons.

Bley, D.C., S. Kaplan, and D.H. Johnson. 1992. The Strengths and Limitations of PSA: Where We Stand. Reliability Engineering and Systems Safety 38. Belfast, Northern Ireland: Elsevier Science Publishers Ltd.

Chess, C., M. Greenberg, and M. Tumuz. 1995. Organizational learning about environmental risk communication. Society and Natural Resources 8: 57–66.

Cramér, Harald. 1946. Mathematical Methods of Statistics. Princeton, N.J.: Princeton University Press.

de Finetti, Bruno. 1974. Theory of Probability: A Critical Introductory Treatment. New York: John Wiley & Sons.

Feller, William. 1968. An Introduction to Probability Theory and Its Applications. 3rd ed. New York: John Wiley & Sons.

Fisher, R.A. 1990. Statistical Methods, Experimental Design, and Scientific Inference. New York: Oxford Science Publications.

Jaynes, E.T. 1983. Confidence intervals versus Bayesian intervals (1976). R.D. Rosenkrantz (ed). E.T. Jaynes: Papers on Probability, Statistics and Statistical Physics, pp. 149–209. Dordrecht, Holland: D. Reidel.

Jaynes, E.T. 1996. Probability Theory: The Logic of Science. Unpublished manuscript. St. Louis, Mo.: Washington University. ftp://bayes.wustl.edu/pub/Jaynes/book.probability.theory/pdf

Jeffreys, Harold. 1961. Theory of Probability. 3rd ed. Oxford, U.K.: Clarendon Press.

Kaplan, S., and B.J. Garrick. 1979. On the use of a Bayesian reasoning in safety and reliability decisions—Three examples. Nuclear Technology 44 (July): 231–245.

Kaplan, S., G. Apostolakis, B.J. Garrick, D.C. Bley, and K. Woodward. 1981. Methodology for Probabilistic Risk Assessment of Nuclear Power Plants (PLG-0209). Newport Beach, Calif.: Pickard, Lowe and Garrick, Inc.

Krüger, L., L.J. Daston, and M. Heidelberger, eds. 1990. The Probabilistic Revolution. Vol. 1. Ideas in History. Vol. 2. Ideas in the Sciences. Cambridge, Mass.: MIT Press.

Lane, David A. 1980. Fisher, Jeffreys, and the Nature of Probability. In S.E. Fienberg and D.V. Hinkley, eds. R.A. Fisher: An Appreciation, pp. 148–160. Berlin, Heidelberg: Springer-Verlag.

Lindley, D.V. 1965. Introduction to Probability and Statistics from a Bayesian Viewpoint: Probability and Inference. 2 vols. New York: Cambridge University Press.

MITRETEK Systems. 1996. Report of the Risk Assessment Expert Panel on the Tooele Chemical Agent Disposal Facility Quantitative Risk Assessment. McLean, Va.: MITRETEK Systems.

Mosleh, A., N. Siu, C. Smidts, C. Liu., eds. 1994. Proceedings of Workshop I in Advanced Topics in Risk and Reliability Analysis—Model Uncertainty: Its Characterization and Quantification. NUREG/CP-0138. Washington, D.C.: U.S. Nuclear Regulatory Commission.

NRC (National Research Council). 1983. Risk Assessment in the Federal Government: Managing the Process. National Research Council. Committee on the Institutional Means for Assessment of Risks to Public Health. Washington, D.C.: National Academy Press.

NRC. 1989. Improving Risk Communication. National Research Council. Committee on Risk Perception and Communication. Washington, D.C.: National Academy Press.

NRC. 1994. Science and Judgment in Risk Assessment. National Research Council. Committee on Risk Assessment of Hazardous Air Pollutants. Washington, D.C.: National Academy Press.

NRC. 1996a. Public Involvement and the Army Chemical Stockpile Disposal Program. National Research Council. Committee on Review and Evaluation of the Army Chemical Stockpile Disposal Program. Washington, D.C.: National Academy Press.

NRC. 1996b. Understanding Risk: Informing Decisions in a Democratic Society. National Research Council. Committee on Risk Characterization. Washington, D.C.: National Academy Press.

Ochsner, M., C. Chess, and M. Greenberg. 1995. Case Study: Du Pont's Edgemoor Facility. Pollution Prevention Review 6(1): 65–74.

OTA (Office of Technology Assessment). 1995. Environmental Policy Tools. Washington, D.C.: U.S. Government Printing Office.

Piller, Charles. 1991. The Fail-Safe Society: Community Defiance and the End of American Technological Optimism. Los Angeles, Calif.: University of California Press.

Savage, Leonard J. 1954. The Foundations of Statistics. New York: John Wiley & Sons.

U.S. Army. 1996. Tooele Chemical Agent Disposal Facility Quantitative Risk Assessment. SAIC-96/2600. Aberdeen Proving Ground, Md.: U.S. Army Program Manager for Chemical Demilitarization.

U.S. NRC (U.S. Nuclear Regulatory Commission). 1975. Reactor Safety Study. WASH-1400, NUREG-75/014. Washington, D.C.: U.S. Nuclear Regulatory Commission.

U.S. NRC. 1995. Federal Register 60FR42622. September 29, 1995. PS-AD-35 to PS-AD-42. Washington, D.C.: U.S. Nuclear Regulatory Commission.

Utah DSHW (Division of Solid and Hazardous Waste). 1996. Tooele Chemical Demilitarization Facility Screening Risk Assessment. EPA I.D. No. UT5210090002. Salt Lake City, Utah: Utah Department of Environmental Quality.

Vesely, W.E., J.W. Chang, E.J. Butcher, and A.C. Thadani. 1995. Risk Management Strategies—Qualitative and Quantitative Approaches. Pp. 869-876 in Proceedings of the International Conference on Probabilistic Safety Assessment Methodology and Applications, Seoul, Korea, November 26–30, 1995. Taejon, Republic of Korea: Korea Atomic Energy Research Institute.

von Mises, R. 1957. Probability, Statistics and Truth. New York: Dover Publications.

APPENDIX B

Risk Assessment Expert Panel on the Tooele Chemical Agent Disposal Facility Quantitative Risk Assessment

The Risk Assessment Expert Panel on the Tooele Chemical Agent Disposal Facility Quantitative Risk Assessment (the Expert Panel), was established to provide an ongoing independent review of the DCD/TOCDF QRA. The Expert Panel is a group of five experts who were brought together under contract with MITRETEK Systems, Inc., and who operate independently of project management. Three of the panel members have extensive QRA experience, primarily in the field of nuclear reactor safety, with additional experience in the analysis of aerospace and chemical process facilities. One panel member is a member of the U.S. Nuclear Regulatory Commission's Advisory Committee on Reactor Safeguards. Another is a combustion expert from Brigham Young University in Salt Lake City, Utah, who also provides some degree of local perspective for the panel. Two are professors of engineering at major universities. Two are chemical engineers with process safety experience. All five have extensive professional experience and are consultants for major organizations. Biographical information on members of the Expert Panel follows:

George Apostolakis is a professor of nuclear engineering at the Massachusetts Institute of Technology. His research interests include mathematical methods for risk and reliability assessment of complex technological systems; uncertainty analysis; decision analysis; fire risk assessment; human reliability models; expert systems; the application of probabilistic models to safety and reliability analyses of nuclear reactors; chemical process systems, space systems, and the control of hazardous substances.

Dr. Apostolakis has served as a consultant to many organizations. Currently, he is a member of the Senior Seismic Hazards Analysis Committee of the Department of Energy, Nuclear Regulatory Commission, and the Electric Power Research Institute. He is a member of the Sandia National Laboratories Probabilistic Risk Assessment (PRA) Working Group and was a senior PRA advisor for U.S. Department of Energy Office of the Director, New Production Reactors. His recent work includes the development of an integrated approach to incorporating organizational performance into probabilistic safety assessment methodology and formal methods for incorporating expert judgment into risk assessments.

Dr. Apostolakis has received many honors and special recognition. He is a fellow of the Society for Risk Analysis and the American Nuclear Society. He served as honorary chairman of the American Nuclear Society Topical Meeting on PSA in 1993. In 1991, he received an Outstanding Service Award from the Society for Risk Analysis.

Formerly, Dr. Apostolakis was a professor in the School of Engineering and Applied Science at the University of California at Los Angeles. His Ph.D., from the California Institute of Technology, is in engineering science and applied mathematics.

Robert J. Budnitz has been involved with the safety of both nuclear and chemical installations for many years. From 1978 to 1980, he was deputy director, then director of the Nuclear Regulatory Commission's (U.S. NRC) Office of Nuclear Regulatory Research. In 1981, Dr. Budnitz formed a private consulting firm,

Future Resources Associates, Inc. His clients have included U.S. government organizations, including the U.S. NRC, the U.S. Department of Energy (DOE), the Environmental Protection Agency, the U.S. Army, and the National Science Foundation, as well as foreign government organizations, such as the International Atomic Energy Agency and the Organization for Economic Cooperation and Development.

Dr. Budnitz has served as chairman and committee member of many professional groups. Currently, he chairs the National Research Council Committee on Remediation of Buried and Tank Wastes. He is chairman of DOE's Oversight Panel for the Yucca Mountain Seismic Hazards Evaluation and chairman of the Westinghouse Savannah River Company's Senior Seismic Advisory Panel. He is a member of DOE's Expert Panel on Aircraft Crash Risk Analysis Methodology and a member of the Sandia National Laboratories peer review team for the Waste Isolation Pilot Plant performance assessment. Dr. Budnitz is a U.S. representative on the European Bank for Reconstruction and Development Safety Review Group for the nuclear safety of Soviet designed reactors. In 1987, Dr. Budnitz served as a member of the review panel organized by General Atomics for the U.S. Army Chemical Munitions Disposal Risk Assessment.

Dr. Budnitz has been prominent in the field of nuclear reactor safety assessment, including probabilistic risk assessment. In recent years, Dr. Budnitz's research has concentrated on the analysis of external accident initiators (e.g., earthquakes, floods, winds, and aircraft). Dr. Budnitz has published numerous papers and reports, among them the principal invited paper on external events at the International Topical Meeting on Probabilistic Safety Assessment, which was cited as the "Outstanding Paper." Dr. Budnitz has a B.A. from Yale University and an M.A. and Ph.D. from Harvard University, all in physics.

Paul O. Hedman is a professor of chemical engineering at Brigham Young University. His interests include chemical and jet propulsion, combustion and gasification, laser instrumentation, fossil energy, and reacting flows. Dr. Hedman has led research in these areas under contracts with several organizations, including the U.S. Department of Energy, the Electric Power Research Institute, the National Science Foundation, and the Tennessee Valley Authority. He has taught courses on principles of chemical processes, heat and mass transfer, thermodynamics, combustion, and several other subjects. He has also published numerous technical papers and reports. While on professional development leave at United Technologies Research Center, Dr. Hedman conducted combustion research in high pressure diffusion and premixed gaseous flames. He has also held fellowships at Wright-Patterson Air Force Base, where he made combustion measurements using laser diagnostics on a simulated jet engine combustor.

Dr. Hedman has been a consultant to the Environmental Protection Agency's Science Advisory Board; the National Bureau of Standards; Occidental Research Corporation; Lockheed Research; Jaycor, Inc.; Atlantic Research Corporation; and Utah Power and Light. He is a member of the Combustion Institute and three honor societies. He is also an independent member of the American Flame Research Committee of the International Flame Research Foundation.

Dr. Hedman's previous professional experience includes four years at Marquardt Corporation, six years at Thiokol Chemical Corporation, three years at Lockheed Propulsion Company, two years with Tetra Tech, Inc., and two years with the U.S. Energy Research and Development Administration. Dr. Hedman received his B.S. degree in mechanical engineering from the University of Utah and his Ph.D. in chemical engineering from Brigham Young University.

Gareth W. Parry is a senior advisor on probabilistic risk assessment (PRA) in the Office of Nuclear Reactor Regulation at the U. S. Nuclear Regulatory Commission. For the major part of this review, he was a project manager in the Energy Risk and Reliability Department of Halliburton NUS. Throughout his career, Dr. Parry has provided expertise in several key areas of PRA, such as data analysis and parameter estimation, common cause failure analysis, external hazard analysis, human reliability analysis, and uncertainty analysis. He has written numerous publications in these areas.

Some of Dr. Parry's recent projects include: managing the individual plant examination for external events (IPEEE) being performed by NUS for three nuclear stations; participating in the development of methods for human reliability analysis sponsored by the Nuclear

Regulatory Commission (U.S. NRC); and making improvements to the systematic human action reliability procedure (SHARP).

Dr. Parry has done extensive development work on the analysis of common cause failures (CCFS) and authored U.S. NRC/Electric Power Research Institute and International Atomic Energy Agency procedural guides for CCF analysis. He has also performed data development and uncertainty analyses in support of PRAs for numerous nuclear power plants. He was a principal author of the chapter on Uncertainty and Sensitivity Analysis in the U.S. NRC PRA Procedures Guide.

Prior to his employment with NUS, Dr. Parry worked for the United Kingdom's Atomic Energy Authority. He also has carried out research in theoretical high energy physics at the University of Durham and the International Centre of Theoretical Physics, in Trieste, Italy. Dr. Parry has a Ph.D in theoretical physics from the Imperial College, London University.

Richard W. Prugh is president of Process Safety Engineering, Inc., (PSE), of Wilmington, Delaware. As a chemical process safety consultant, he has conducted many process safety studies as well as authored numerous papers. His specialties include chemical process safety analysis, explosion hazards analysis, toxic vapor cloud analysis, safety assessments, fire protection, and electrical hazards analysis.

Mr. Prugh has been with PSE (formerly known as Hazard Reduction Engineering) since 1985. During this time he was also a consultant to E.I. duPont de Nemours and CONDUX, Inc., and has been a part-time staff member of the American Institute of Chemical Engineers Center for Chemical Process Safety. From 1955 to 1985, Mr. Prugh was employed by E.I. duPont de Nemours. From 1952 to 1954, he served in the U.S. Air Force as a second lieutenant in the Air Rescue Service.

Mr. Prugh has recently conducted several process safety studies, including an Occupational Safety and Health Administration (OSHA) and Environmental Protection Agency (EPA) HAZOP study for a chemical plant in North Carolina; a failure modes and effects analysis for a chemical plant in New Jersey; an overpressure protection design study for an electrical equipment manufacturer in New York; and a pre-startup safety review for OSHA/EPA for a chemical plant in Arkansas.

Mr. Prugh has an M.S. degree in chemical engineering from Stevens Institute of Technology. He has undertaken additional studies in the areas of nuclear engineering (Massachusetts Institute of Technology), biomedical engineering (Drexel Institute of Technology), business law (Temple University) and chemical engineering (University of Delaware). Mr. Prugh is a certified quantitative consequence analyst, a certified safety professional, and a hazardous materials first responder. He is also certified as a professional engineer in New Jersey, Delaware, Pennsylvania, and California.

APPENDIX C

Reports of the Committee on Review and Evaluation of the Army Chemical Stockpile Disposal Program (Stockpile Committee)

Comments on Operational Verification Test and Evaluation Master Plan for the Johnston Atoll Chemical Agent Disposal System (JACADS). National Research Council. Board on Army Science and Technology (1989).

Demilitarization of Chemical Weapons: Cryofracture. National Research Council. Board on Army Science and Technology (1989).

Demilitarization of Chemical Weapons: On-Site Handling of Munitions. National Research Council. Board on Army Science and Technology (1989).

Letter Report Commenting on Proposed Cryofracture Program Testing. National Research Council. Board on Army Science and Technology (1991).

Letter Report on Review of the MITRE Report: Evaluation of the GB Rocket Campaign: Johnston Atoll Chemical Agent Disposal System Operational Verification Testing, dated May 1991. National Research Council. Board on Army Science and Technology (1991).

Letter Report on Siting of a Cryofracture Chemical Stockpile Facility. National Research Council. Board on Army Science and Technology (1991).

Letter Report on Workshop on the Pollution Abatement System of the Chemical Agent Demilitarization System. National Research Council. Board on Army Science and Technology (1991).

Letter Report on Review of the Choice and Status of Incineration for Destruction of the Chemical Stockpile. National Research Council. Board on Army Science and Technology (1992).

Evaluation of the Johnston Atoll Chemical Agent Disposal System Operational Verification Testing: Part I. National Research Council. Board on Army Science and Technology (1993)

Letter Report to Recommend Specific Actions to Further Enhance the CSDP [Chemical Stockpile Disposal Program] Risk Management Process. National Research Council. Board on Army Science and Technology (1993).

Evaluation of the Johnston Atoll Chemical Agent Disposal System Operational Verification Testing: Part II. National Academy Press (1994).

Recommendations for the Disposal of Chemical Agents and Munitions. National Academy Press (1994).

Review of Monitoring Activities Within the Army Chemical Stockpile Disposal Program. National Academy Press (1994).

Evaluation of the Army's Draft Assessment Criteria to Aid in the Selection of Alternative Technologies for Chemical Demilitarization. National Academy Press (1995).

Letter Report: Public Involvement and the Army Chemical Stockpile Disposal Program. National Research Council. Board on Army Science and Technology (1996).

Review of Systemization of the Tooele Chemical Agent Disposal Facility. National Academy Press (1996).

APPENDIX D

Biographical Sketches of Committee Members

Richard S. Magee *(chair),* is a professor in the Department of Mechanical Engineering and the Department of Chemical Engineering, Chemistry, and Environmental Science and executive director of the Center for Environmental Engineering and Science at New Jersey Institute of Technology (NJIT). He also directs the U.S. Environmental Protection Agency's Northeast Hazardous Substance Research Center, as well as the Hazardous Substance Management Research Center, which is jointly sponsored by the National Science Foundation and the New Jersey Commission on Science and Technology, both headquartered at NJIT. He is a fellow of the American Society of Mechanical Engineers (ASME) and a diplomate of the American Academy of Environmental Engineers. Dr. Magee's research expertise is in combustion, with a major focus on the incineration of municipal and industrial wastes. He has served as vice chairman of the ASME Research Committee on Industrial and Municipal Wastes and as a member of the United Nations Special Commission (under Security Council Resolution 687) Advisory Panel on Destruction of Iraq's Chemical Weapons Capabilities. He presently serves as a member of the North Atlantic Treaty Organization Science Committee's Priority Area Panel on disarmament technologies. He recently served as chair of the National Research Council Panel on Review and Evaluation of Alternative Chemical Disposal Technologies.

Elisabeth M. Drake *(vice chair),* a member of the National Academy of Engineering, is the associate director of the Massachusetts Institute of Technology Energy Laboratory. A chemical engineer with interest and experience in risk management and technology associated with the transport, processing, storage, and disposal of hazardous materials, as well as with chemical engineering process design and control systems, Dr. Drake has a special interest in the interactions between technology and the environment. She has served extensively as both a consultant to government and industry and as a professor of chemical engineering. She has been very active with the American Institute of Chemical Engineers, in particular with the Center for Chemical Process Safety. She belongs to a number of environmental organizations, including the Audubon Society, the Sierra Club, and Greenpeace.

Dennis C. Bley is president of Buttonwood Consulting, Inc., and a principal of The WreathWood Group, a joint venture supporting multidisciplinary research in human reliability. He has more than 25 years of experience in nuclear and electrical engineering, reliability and availability analysis, plant and human modeling for risk assessment, diagnostic system development, and technical management. Dr. Bley has a Ph.D. in nuclear engineering from the Massachusetts Institute of Technology and is a registered professional engineer in the State of California. Dr. Bley has served on a number of technical review panels for U.S. Nuclear Regulatory Commission and Department of Energy programs and is a frequent lecturer in short courses for universities, industries, and government agencies. He is active in many professional organizations and on the board of directors of the International Association for Probabilistic Safety Assessment and Management. He has published extensively on subjects related to risk assessment, and his current research interests include applying risk analysis to diverse technological systems, modeling uncertainties in risk analysis and risk management, technical risk communication, and human reliability analysis.

Gene H. Dyer graduated with a B.S. degree in chemistry, mathematics, and physics from the University of Nebraska. Over a 12-year period he worked for General Electric as a process engineer, the U.S. Navy as a research and development project engineer, and the U.S. Atomic Energy Commission as a project engineer. In 1963, he then began a more than 20-year career with the Bechtel Corporation. First a consultant on advanced nuclear power plants and later a program supervisor for nuclear facilities, he served as manager of the Process and Environmental Department from 1969 to 1983. This department provided engineering services related to research and development projects, including technology probes, environmental assessment, air pollution control, water pollution control, process development, nuclear fuel process development, and regional planning. He culminated his career at Bechtel as a senior staff consultant for several years, responsible for identifying and evaluating new technologies and managing their development and testing for practical applications. He is a member of the American Institute of Chemical Engineers and a registered professional engineer. He recently served as a member of the National Research Council Committee on Alternative Chemical Demilitarization Technologies.

Vincent E. Falter spent more than 34 years in the U.S. Army, about half of that time dealing with nuclear weapons. Major General Falter was director of nuclear and chemical warfare on the Army Staff and was the single point of contact for all chemical operations for the U.S. Department of Defense. He was assigned responsibility for all chemical weapons and for initiating their destruction. He initiated the funding for the Johnston Atoll Chemical Agent Disposal System. He retired from the Army approximately eight years ago. Since then, he has been a national security research analyst and consultant for numerous corporations. He was a member of the Joint Strategic Targeting Planning Staff at the Strategic Air Command; the Scientific Advisory Committee for Nuclear Weapons Effects; and was the U.S. Department of Defense representative for two rounds of the chemical disarmament talks.

J. Robert Gibson is the assistant director of the Haskell Laboratory, E.I. duPont de Nemours & Company, and an adjunct associate professor of marine studies at the University of Delaware. After receiving his Ph.D. in physiology from Mississippi State University, Dr. Gibson specialized in toxicology for more than 20 years. He was certified by the American Board of Toxicology and has written numerous publications.

Michael R. Greenberg is a professor in the Department of Urban Studies and Community Health at Rutgers, The State University of New Jersey, and is an adjunct professor of environmental and community medicine at the Robert Wood Johnson Medical School. His principal research and teaching interests include urbanization, industrialization, and environmental health policy. Dr. Greenberg holds a B.A. in mathematics and history, an M.A. in urban geography, and a Ph.D. in environmental and medical geography.

Charles E. Kolb is president and chief executive officer of Aerodyne Research, Inc. At Aerodyne since 1971, his principal research interests have included atmospheric and environmental chemistry, combustion chemistry, materials chemistry, and the chemical physics of rocket and aircraft exhaust plumes. He has served on several National Aeronautics and Space Administration panels dealing with atmospheric chemistry and global change, as well as on five National Research Council committees and boards dealing with environmental issues. He is currently atmospheric sciences editor for the American Geophysical Union journal, *Geophysical Research Letters*, and recently received the Award for Creative Advances in Environmental Science and Technology from the American Chemical Society.

David S. Kosson graduated with a bachelor of science degree in chemical engineering, a master's degree in chemical and biochemical engineering, and a doctorate in chemical and biochemical engineering from Rutgers, The State University of New Jersey. He joined the faculty at Rutgers in 1986 and was made an associate professor with tenure in 1990 and a full professor in 1996. He teaches graduate and undergraduate chemical engineering and environmental engineering courses. In addition, he carries out research for the Department of Chemical and Biochemical Engineering, where considerable work is under way in developing microbial, chemical, and physical treatment methods for hazardous waste. He is responsible for project planning and coordination, from basic research through full-scale design and implementation. He has published extensively in the fields of chemical engineering, waste management and treatment, and contaminant fate and transport in soils and groundwater. Dr. Kosson has served on several Environmental Protection Agency advisory panels involved in waste research and is the director of the Physical Treatment Division of the Hazardous Substances Management Research Center in New Jersey. He is a member of the American Institute of Chemical Engineers and recently served as a member of the National Research Council Committee on Alternative Chemical Demilitarization Technologies.

Walter G. May graduated with a bachelor of science degree in chemical engineering and master of science degree in chemistry from the University of Saskatchewan and with a doctor of science degree in chemical engineering from the Massachusetts Institute of Technology. He joined the faculty of the University of Saskatchewan as a professor of chemical engineering in 1943. In 1948, he began a distinguished career with Exxon Research and Engineering Company, where he was a senior science advisor from 1976 to 1983. He was professor of chemical engineering at the University of Illinois from 1983 until his retirement in 1991. There he conducted courses in process design, thermodynamics, chemical reactor design, separation processes, and industrial chemistry and stoichiometry. Dr. May has published extensively, served on the editorial boards of *Chemical Engineering Reviews* and *Chemical Engineering Progress*, and has obtained numerous patents in his field. He is a member of the National Academy of Engineering and a fellow of the American Institute of Chemical Engineers, and he has received special awards from the American Institute of Chemical Engineers and the American Society of Mechanical Engineers. He has a particular interest in separations research. He is a registered professional engineer in the state of Illinois. He recently served as a member of the National Research Council Committee on Alternative Chemical Demilitarization Technologies and the Committee on Decontamination and Decomissioning of the Uranium Diffusion Plants.

Alvin H. Mushkatel, professor in the School of Planning and Landscape Architecture, Arizona State University, is an expert in emergency management risk perceptions. His research interests include emergency management, natural and technological hazards policy, and environmental policy. He has been a member of the National Research Council Committee on Earthquake Engineering, the Committee on Decontamination and Decommissioning of Uranium Enrichment Facilities and the Panel on Review and Evaluation of Alternative Chemical Disposal Technologies. His most recent research focuses on intergovernmental policy conflicts involving high-level nuclear waste disposal and the role of citizens in technological policy decision-making processes. He has published extensively on issues related to siting controversies.

Peter J. Niemiec, a partner in the law firm of Greenberg, Glusker, Fields, Claman & Machtinger, LLP, in Los Angeles, California, is an expert in environmental law and regulations. His work in the private sector has focused on the regulation of, and liability arising from, hazardous materials, including extensive work on Superfund issues. Mr. Niemiec has also represented federal and state environmental agencies, where he was involved in the development of national enforcement policies and permitting and enforcement for major industrial facilities and landfill disposal sites. He has also been an adjunct professor at the Indiana School of Law (Indianapolis), where he taught environmental law. He has published several articles on the availability of private remedies for environmental cleanup.

Dr. George W. Parshall was director of chemical science in the Central Research and Development Department of the DuPont Company from 1979 until his retirement at the end of 1992. He began his career with DuPont in 1954 and later supervised a group doing research in inorganic chemistry and catalysis. Since retirement, he has been a consultant for DuPont and has participated in advisory activities through the National

Research Council for the Board on Physics and Astronomy, the Committee on Environmental Management Technologies, and the Panel on Review and Evaluation of Alternative Chemical Disposal Technologies, in addition to the Stockpile Committee. Dr. Parshall is a member of the National Academy of Sciences and the American Academy of Arts and Sciences.

William Tumas is currently the group leader for the Waste Treatment and Minimization Science and Technology Group at Los Alamos National Laboratory. He is a senior chemist known primarily for his work in science and engineering research on waste treatment and minimization. His work has included research and technology development for industrial waste applications and environmental restoration for the U.S. Department of Energy. At Los Alamos, he has studied supercritical fluids, oxidation, and organic transformations. He has written numerous papers and is a member of several professional organizations. In addition, Dr. Tumas was recently a member of the National Research Council Panel on Review and Evaluation of Alternative Chemical Disposal Technologies.

Jya-Syin Wu, a systems safety engineer at Hughes Information Technology Systems, is currently working on the system safety analysis of the Wide Area Augmentation System, a means of augmenting information from the Global Positioning System for air navigation for the Federal Aviation Administration. Dr. Wu holds a Ph.D. in nuclear science and engineering from the University of California, Los Angeles. She has more than 15 years of experience working on probabilistic risk assessments for nuclear power plants and has published many papers in major technical journals. Her recent interests have been focused on the risk assessment of complex engineering systems and safety-critical software systems.